T0348581

VOLUME EIGHTY TWO

Advances in
APPLIED MICROBIOLOGY

VOLUME EIGHTY TWO

Advances in
APPLIED MICROBIOLOGY

Edited by

SIMA SARIASLANI
Wilmington, Delaware, USA

GEOFFREY M. GADD
Dundee, Scotland, UK

AMSTERDAM • BOSTON • HEIDELBERG • LONDON
NEW YORK • OXFORD • PARIS • SAN DIEGO
SAN FRANCISCO • SINGAPORE • SYDNEY • TOKYO
Academic Press is an imprint of Elsevier

Academic Press is an imprint of Elsevier
225 Wyman Street, Waltham, MA 02451, USA
525 B Street, Suite 1800, San Diego, CA 92101-4495, USA
Radarweg 29, PO Box 211, 1000 AE Amsterdam, The Netherlands
The Boulevard, Langford Lane, Kidlington, Oxford, OX5 1GB, UK
32 Jamestown Road, London NW1 7BY, UK

First edition 2013

ISBN: 978-0-12-407679-2
ISSN: 0065-2164

For information on all Academic Press publications
visit our website at store.elsevier.com

Working together to grow
libraries in developing countries
www.elsevier.com | www.bookaid.org | www.sabre.org

ELSEVIER   BOOK AID
             International   Sabre Foundation

CONTENTS

4. Recombinant Production of Spider Silk Proteins   **115**

Aniela Heidebrecht, Thomas Scheibel

5. Mechanisms of Immune Evasion in Leishmaniasis   **155**

Gaurav Gupta, Steve Oghumu, Abhay R. Satoskar

CONTRIBUTORS

Ahmed M. Abdel-Hamid
Energy Biosciences Institute, University of Illinois, IL, USA; Institute for Genomic Biology, University of Illinois, IL, USA

Isaac K.O. Cann
Energy Biosciences Institute, University of Illinois, IL, USA; Institute for Genomic Biology, University of Illinois, IL, USA; Department of Animal Sciences, University of Illinois, IL, USA; Department of Microbiology, University of Illinois, IL, USA

James E. Graham
Department of Microbiology and Immunology, and Department of Biology, University of Louisville, Louisville, KY, USA

Gaurav Gupta
Department of Pathology, The Ohio State University Medical Center, Columbus, OH, USA

Aniela Heidebrecht
Department of Biomaterials, University of Bayreuth, Universitätsstr. 30, 95440 Bayreuth, Germany

Steve Oghumu
Department of Pathology, The Ohio State University Medical Center, Columbus, OH, USA; Department of Oral Biology, The Ohio State University College of Dentistry, Columbus, OH, USA

Chilekampalli A. Reddy
Department of Microbiology and Molecular Genetics, Michigan State University, East Lansing, MI, USA

Ramu S. Saravanan
Department of Microbiology and Molecular Genetics, Michigan State University, East Lansing, MI, USA

Abhay R. Satoskar
Department of Pathology, The Ohio State University Medical Center, Columbus, OH, USA; Department of Microbiology, The Ohio State University, Columbus, OH, USA

Thomas Scheibel
Department of Biomaterials, University of Bayreuth, Universitätsstr. 30, 95440 Bayreuth, Germany

Jose O. Solbiati
Energy Biosciences Institute, University of Illinois, IL, USA; Institute for Genomic Biology, University of Illinois, IL, USA

CHAPTER ONE

Insights into Lignin Degradation and its Potential Industrial Applications

Ahmed M. Abdel-Hamid*,**, Jose O. Solbiati*,**, Isaac K. O. Cann*,**,#,@,1

*Energy Biosciences Institute, University of Illinois, IL, USA
**Institute for Genomic Biology, University of Illinois, IL, USA
#Department of Animal Sciences, University of Illinois, IL, USA
@Department of Microbiology, University of Illinois, IL, USA
1Corresponding author: E-mail: icann@illinois.edu

Contents

Advances in Applied Microbiology, Volume 82
ISSN 0065-2164, http://dx.doi.org/10.1016/B978-0-12-407679-2.00001-6

Abstract

Lignocellulose is an abundant biomass that provides an alternative source for the production of renewable fuels and chemicals. The depolymerization of the carbohydrate polymers in lignocellulosic biomass is hindered by lignin, which is recalcitrant to chemical and biological degradation due to its complex chemical structure and linkage heterogeneity. The role of fungi in delignification due to the production of extracellular oxidative enzymes has been studied more extensively than that of bacteria. The two major groups of enzymes that are involved in lignin degradation are heme peroxidases and laccases. Lignin-degrading peroxidases include lignin peroxidase (LiP), manganese peroxidase (MnP), versatile peroxidase (VP), and dye-decolorizing peroxidase (DyP). LiP, MnP, and VP are class II extracellular fungal peroxidases that belong to the plant and microbial peroxidases superfamily. LiPs are strong oxidants with high-redox potential that oxidize the major non-phenolic structures of lignin. MnP is an Mn-dependent enzyme that catalyzes the oxidation of various phenolic substrates but is not capable of oxidizing the more recalcitrant non-phenolic lignin. VP enzymes combine the catalytic activities of both MnP and LiP and are able to oxidize Mn^{2+} like MnP, and non-phenolic compounds like LiP. DyPs occur in both fungi and bacteria and are members of a new superfamily of heme peroxidases called DyPs. DyP enzymes oxidize high-redox potential anthraquinone dyes and were recently reported to oxidize lignin model compounds. The second major group of lignin-degrading enzymes, laccases, are found in plants, fungi, and bacteria and belong to the multicopper oxidase superfamily. They catalyze a one-electron oxidation with the concomitant four-electron reduction of molecular oxygen to water. Fungal laccases can oxidize phenolic lignin model compounds and have higher redox potential than bacterial laccases. In the presence of redox mediators, fungal laccases can oxidize non-phenolic lignin model compounds. In addition to the peroxidases and laccases, fungi produce other accessory oxidases such as aryl-alcohol oxidase and the glyoxal oxidase that generate the hydrogen peroxide required by the peroxidases. Lignin-degrading enzymes have attracted the attention for their valuable biotechnological applications especially in the pretreatment of recalcitrant lignocellulosic biomass for biofuel production. The use of lignin-degrading enzymes has been studied in various applications such as paper industry, textile industry, wastewater treatment and the degradation of herbicides.

1. LIGNOCELLULOSIC BIOMASS

Lignocellulose is the most abundant renewable biomass on earth. It has long been recognized as an alternative source for producing renewable fuels and chemicals. Lignocellulosic biomass is primarily composed of the two carbohydrate polymers, cellulose and hemicellulose, and the non–carbohydrate phenolic polymer, lignin. Lignin binds to cellulose fibers to harden and strengthen the plant cell walls. Cellulose is the main structural polysaccharide of the primary plant cell wall, and it accounts for 30–50 % of the dry weight of lignocellulose (Foyle, Jennings, & Mulcahy, 2007; Harris & Debolt, 2010), and it consists of linear chains of β(1→4) linked D-glucose units (Updegraff, 1969). Cellulose obtained from non-food energy crops can be degraded by the action of microbial cellulases to glucose monomers, with subsequent conversion to biofuels or other value-added chemicals. The second polysaccharide component of lignocellulose is hemicellulose, which accounts for 15–30% of the plant cell wall. Hemicelluloses are imbedded in the plant cell walls, and one of their main functions is to bind cellulose microfibrils to strengthen the cell wall. Unlike cellulose, hemicellulose has a random and amorphous structure, which is composed of several heteropolymers including xylan, glucuronoxylan, arabinoxylan, glucomannan, and xyloglucan (Scheller & Ulvskov, 2010). Hemicelluloses can be hydrolyzed by dilute acid or base as well as several microbial hemicellulases, a complex set of enzymes with components that remove the side chains and others that attack the backbone randomly to release oligosaccharides that are subsequently degraded to simple sugars. The third main component of lignocellulose is lignin, which accounts for 15–30% of lignocellulose dry mass. Lignin is found in all vascular plants, representing the second most abundant carbon source after cellulose (Boerjan, Ralph, & Baucher, 2003). Lignin provides the cell walls and tissues of vascular plants with strength and rigidity and protects the cell wall against microbial degradation of the structural polysaccharides.

2. CHEMICAL STRUCTURE OF LIGNIN

Unlike cellulose and hemicellulose, lignin is a non-carbohydrate aromatic heteropolymer that is derived from the oxidative coupling of three different phenylpropane building blocks (monolignols): p-coumaryl alcohol, coniferyl alcohol, and sinapyl alcohol. The corresponding phenylpropanoid monomeric units in the lignin polymer are known as p-hydroxyphenyl (H), guaiacyl (G), and syringyl (S) units, respectively (Faix, 1991; Kleinert & Barth, 2008; Wong, 2009). The chemical structure of the primary lignin-building blocks and their corresponding lignin polymer monomeric units is illustrated in Fig. 1.1.

The abundance of the three monomers varies in lignin according to the plant species and plant tissue. Lignin from grasses is built up by the three monomeric units (G, S, and H), lignin from hardwood contains roughly equal amounts of (G) and (S) units, and lignin from softwood is

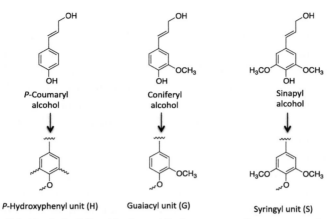

Figure 1.1 Lignin-building blocks and their corresponding monomeric units.

composed mainly of (G) units accounting for up to 90% of the total units (Faix, 1991).

Lignin monomers polymerize by the coupling of two radicals forming a dimer by β-O-4-aryl ether bonds. Further polymerization of monomeric radicals with dimers, trimers, and oligomers progresses producing a complex branched polymer (Amen-Chen, Pakdel, & Roy, 2001; Pandey & Kim, 2011; Reale, Di Tullio, Spreti, & De Angelis, 2004). Linkages in lignin can be divided into two broad groups: C–O ether linkages and C–C linkages, including β-O-4, α-O-4, 5-5, β-5, 4-O-5, β-1, and β–β, of which β-O-4 is the most common linkage in lignin (Parthasarathi, Romero, Redondo, & Gnanakaran, 2011). The proportion of these linkages varies according to plant species. However, typically more than two-thirds of the linkages in lignin are ether linkages (Dorrestijn, Laarhoven, Arends, & Mulder, 2000). Figure 1.2 illustrates the most common linkages found in lignin.

3. LIGNIN-DEGRADING MICROORGANISMS

Lignin is chemically recalcitrant to degradation by most organisms because of its complex structure. Due to linkage heterogeneity, lignin cannot be cleaved by hydrolytic enzymes as observed for degradation of the other cell wall components. Lignin breakdown has been extensively studied in wood-rotting organisms, especially white-rot basidiomycetes (Boyle, Kropp, & Reid, 1992; Leonowicz et al., 1999), and here we summarize some of the pertinent knowledge in this area.

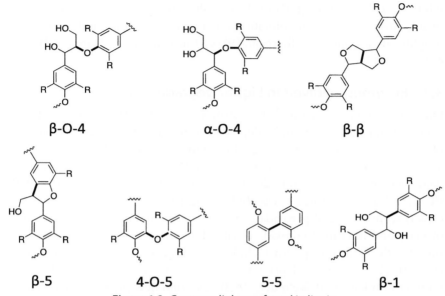

Figure 1.2 Common linkages found in lignin.

3.1. Lignin Degradation by Fungi

3.1.1. White-rot Fungi

White-rot fungi are the most effective for delignification due to production of ligninolytic extracellular oxidative enzymes. Lignin degradation by several white-rot fungi, such as *Phanerochaete chrysosporium, Pleurotus ostreatus, Coriolus versicolor, Cyathus stercoreus*, and *Ceriporiopsis subvermispora*, have been studied (Martinez et al., 2004; Ruttimann-Johnson, Salas, Vicuna, & Kirk, 1993; Wan & Li, 2012). White-rot fungi degrade lignin leaving decayed wood whitish in color and fibrous in texture. Some white-rot fungi such as *C. subvermispora, Phellinus pini, Phlebia* spp., and *Pleurotus* spp. delignify wood by preferentially attacking lignin more readily than hemicellulose and cellulose, leaving enriched cellulose. However, other white-rot fungi such as *Trametes versicolor, Heterobasidion annosum*, and *Irpex lacteus* degrade the cell wall components simultaneously (Wong, 2009).

3.1.2. Brown-rot Fungi

Another group of wood-decaying fungi is the brown-rot fungi such as *Gloeophyllum trabeum, Laetiporus portentosus*, and *Fomitopsis lilacinogilva*, which grow mainly on conifers and represent only 7% of wood-rotting basidiomycetes. Unlike white-rot fungi, brown-rot fungi degrade wood polysaccharides while partially modifying lignin. As a result of this type of decay, the wood shrinks, shows a brown discoloration due to oxidized lignin, and cracks into roughly cubical pieces (Gilbertson, 1980; Monrroy, Ortega, Ramírez, Baeza, & Freer, 2011).

3.2. Enzymes Involved in Lignin Degradation

Due to the complex nature of its structure and also the wide variety of chemical linkages in lignin, its degradation requires the synergistic action of several enzymes. Lignin-degrading fungi produce three extracellular heme peroxidases: lignin peroxidase (LiP, EC 1.11.1.14), manganese-dependent peroxidase (MnP, EC 1.11.1.13), and versatile peroxidase (VP, EC 1.11.1.16). The three extracellular fungal LiP, MnP, and VP belong to class II secreted fungal peroxidases, which are members of the superfamily of plant and microbial peroxidases (also known as non-animal peroxidase superfamily). This group also includes class I, prokaryotic peroxidases and class III, classical secreted plant peroxidases (Hofrichter, Ullrich, Pecyna, Liers, & Lundell, 2010; Welinder, 1992). A second new superfamily of heme peroxidases called DyPs (EC 1.11.1.19) has been identified in both fungi and bacteria (Sugano, 2009). DyPs were reported to oxidize high-redox potential dyes as well as the β-O-4 non-phenolic lignin model compounds (Liers, Bobeth, Pecyna, Ullrich, & Hofrichter, 2010). Another major group of enzymes involved in lignin degradation are the copper-containing phenol oxidases (Laccase, EC 1.10.3.2), which are produced by wood-degrading fungi as well as bacteria (Leonowicz et al., 2001). Fungi also enhance the process of lignin degradation by producing several oxidoreductases, which include glyoxal oxidase (EC 1.2.3.5), aryl alcohol oxidase (veratryl alcohol oxidase; EC 1.1.3.7), pyranose 2-oxidase (glucose 1-oxidase; EC 1.1.3.4), cellobiose/quinone oxidoreductase (EC 1.1.5.1), and cellobiose dehydrogenase (EC 1.1.99.18) (Ander & Marzullo, 1997). These enzymes function as supporting enzymes that regulate lignin degradation in white-rot fungi by reducing the methoxy radicals generated by LiP, MnP, and laccase. In addition, glyoxal oxidase, glucose oxidase, and veratryl alcohol oxidase produce hydrogen peroxide (H_2O_2) required by the peroxidases (Ander & Marzullo, 1997). The role of the different enzymes involved in lignin degradation is illustrated in Fig. 1.3.

3.2.1. Class II Fungal Peroxidases

Fungal peroxidases are extracellular heme-containing enzymes that catalyze a number of oxidative reactions and hydroxylations, using H_2O_2 as

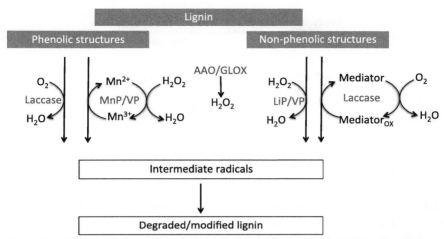

Figure 1.3 *The reaction catalyzed by lignin-degrading enzymes.* LiP: lignin peroxidase, MnP: manganese peroxidase, AAO: aryl alcohol oxidase, GLOX: glyoxal oxidase.

the electron acceptor. Class II peroxidases consists of the secretory fungal peroxidases such as LiP, MnP, and VP.

3.2.1.1. Lignin Peroxidase

LiP (EC 1.11.1.14) catalyzes the H_2O_2-dependent oxidative depolymerization of lignin (Piontek, Smith, & Blodig, 2001; Poulos, Edwards, Wariishi, & Gold, 1993). LiP was first discovered in the extracellular medium of the white-rot fungus, *P. chrysosporium* (Kirk & Farrell, 1987). Various isoforms of LiP exist in *P. chrysosporium* and other white-rot fungi such as *T. versicolor* (Johansson, Welinder, & Nyman, 1993), *Phanerochaete sordida* (Sugiura et al., 2009), *Phlebia radiata* (Moilanen, Lundell, Vares, & Hatakka, 1996), and *Phlebia tremellosa* (Vares, Niemenmaa, & Hatakka, 1994). LiPs are monomeric hemoproteins with molecular masses around 40 kDa, and like classical peroxidases such as the horseradish peroxidase (HRP) their Fe(III) is pentacoordinated to the four heme tetrapyrrole nitrogens and to a histidine residue. LiP has a typical peroxidase catalytic cycle similar to that of HRP. The general mechanism of LiP catalytic cycle is illustrated in Fig. 1.4. The first reaction involves the two–electron oxidation of the native ferric enzyme [Fe(III)] by H_2O_2 to form a compound I oxo-ferryl intermediate (two–electron oxidized form). In the second reaction, compound-I is reduced by a reducing substrate such as non-phenolic aromatic substrate (A) and receives one electron to form compound-II (one–electron oxidized form). In the third step, compound II receives a second electron from the

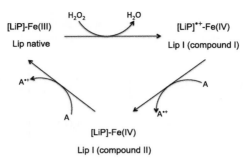

Figure 1.4 The catalytic reaction of LiP.

reduced substrate returning the enzyme to the native ferric oxidation state to complete the oxidation cycle.

LiPs are stronger oxidants with higher redox potential than classical peroxidases since the iron in the porphyrin ring in LiPs is more electron deficient than in classical peroxidases (Millis, Cai, Stankovich, & Tien, 1989). Therefore, LiPs can oxidize aromatic rings that are moderately activated, whereas classical peroxidases act only on strongly activated aromatic substrates. Although, both LiP and HRP can oxidize many phenols and anilines, only LiP can oxidize the major non-phenolic structures of lignin, which constitute up to 90% of the polymer (Martinez et al., 2005). Lip was shown to oxidize a variety of non-phenolic lignin model compounds such as the β-O-4 linkage-type arylglycerol-aryl ethers in the presence of H_2O_2 (Valli, Wariishi, & Gold, 1990). Oxidation of β-O-4 lignin model compounds by LiP involves the initial formation of radical cation via $1e^-$ oxidation, followed by side-chain cleavage, demethylation, intramolecular addition, and rearrangements (Kirk, Tien, & Kersten, 1986; Miki, Renganathan, & Gold, 1986). During the growth of *P. chrysosporium* on lignin, veratryl alcohol (VA) is produced as a metabolite that enhances the LiP activity and consequently the rate of lignin degradation (Lundell et al., 1993; Schoemaker, Lundell, Hatakka, & Piontek, 1994; Tien, Kirk, Bull, & Fee, 1986). The Trp171 residue in LiP has been suggested to play an important role in the oxidation of high-redox potential non-phenolic substrates such as VA through long-range electron transfer (Choinowski, Blodig, Winterhalter, & Piontek, 1999). The ability of LiP to oxidize VA was lost completely by the substitution of the Trp residue with Ser or Phe (Doyle, Blodig, Veitch, Piontek, & Smith, 1998).

VA, a non-phenolic and high-redox potential substrate, is the preferred aromatic electron donor for LiP and is oxidized to veratraldehyde. VA was

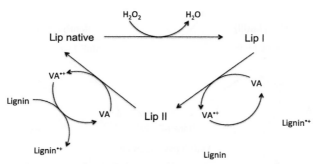

Figure 1.5 VA as a redox mediator in lignin degradation.

suggested to function as a redox mediator for the oxidation of lignin substrates that are not oxidized by LiP (Fig. 1.5). In this model, LiP indirectly oxidizes lignin by oxidizing VA to the corresponding diffusible cation radical (VA·$^+$) that functions as a direct oxidant on lignin (Harvey, Schoemaker, & Palmer, 1986; Wong, 2009). Studies on the measurement of the lifetime of VA cation, to test its ability to function as diffusible oxidant, revealed contradictory results. Some studies suggested that VA·$^+$ is able to function as a diffusible oxidant (Bietti, Baciocchi, & Steenken, 1998; Khindaria, Yamazaki, & Aust, 1996), whereas others reported that it has a very short lifetime and hence will be unable to sustain lignin oxidation (Candeias & Harvey, 1995).

3.2.1.2. Manganese Peroxidase

MnP (EC 1.11.1.13) is a heme-containing glycoprotein that oxidizes the one-electron donor Mn^{2+} to Mn^{3+}, which in turn can oxidize a large number of phenolic substrates. MnP was first discovered in *P. chrysosporium* (Glenn & Gold, 1985; Paszczyński, Huynh, & Crawford, 1985). The production of MnP has been shown in two groups of basidiomycetes including wood-decaying white-rot fungi and soil litter-decomposing fungi (Hofrichter, 2002). White-rot fungi secrete MnP mostly in multiple forms. For example, up to 11 different isoforms of MnP have been described in *C. subvermispora* (Lobos, Larrain, Salas, Cullen, & Vicuna, 1994). The catalytic cycle of MnP is similar to those of other heme-containing peroxidases, such as HRP and LiP; however, it is unique in utilizing Mn^{2+} as the electron donor (Fig. 1.6). The catalytic cycle is initiated by the reaction of the native ferric enzyme and H_2O_2 to form MnP compound I, which is a Fe^{4+}-oxo-porphyrin-radical complex. A monochelated Mn^{2+} ion donates one electron to the porphyrin intermediate to form Compound II and is oxidized to Mn^{3+}. The native enzyme is generated from Compound II in a similar way through

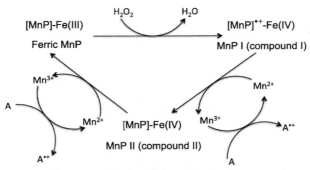

Figure 1.6 The catalytic reaction of MnP.

the donation of one electron from Mn^{2+} to form Mn^{3+}. The chelated Mn^{3+} ions generated by MnP acts as diffusible charge-transfer mediators, allowing the oxidation of various phenolic substrates such as simple phenols, amines, dyes, as well as phenolic lignin model compounds (Hofrichter et al., 2010; Wariishi, Valli, & Gold, 1991). In contrast to LiP, MnP is not capable of oxidizing the more recalcitrant non-phenolic compounds. It has been suggested that white-rot fungi, which produce MnP and laccase but not LiP, may produce mediators to enable MnP to cleave non-phenolic lignin substrates (Reddy, Sridhar, & Gold, 2003). It has also been reported that MnP can cleave non-phenolic lignin structures via the action of small mediators such as thiyl or lipid radicals. MnP from white-rot fungi has been shown to catalyze the peroxidation of unsaturated lipids with subsequent cleavage of non-phenolic lignin model compounds (Bao, Fukushima, Jensen, Moen, & Hammel, 1994; Kapich, Hofrichter, Vares, & Hatakka, 1999).

3.2.1.3. Versatile Peroxidase
VP (EC 1.11.1.16) is a heme-containing ligninolytic peroxidase with hybrid molecular architecture combining different oxidation-active sites (Perez-Boada et al., 2005; Ruiz-Duenas et al., 2009). VP was first described in the white-rot fungus *Pleurotus eryngii* (Martinez, Böckle, Camarero, Guillén, & Martinez, 1996). VP has only been described in species of the genera *Pleurotus* and *Bjerkandera* (Heinfling et al., 1998; Moreira, Almeida-Vara, Malcata, & Duarte, 2007). VP enzymes have the catalytic activities of both MnP and LiP and are able to oxidize Mn^{2+} like MnP, and high-redox potential non-phenolic compounds like LiP. The catalytic cycle of VP combines the cycles of the other fungal peroxidases, LiP and MnP. Unlike the classical MnP, VP is able to catalyze the Mn^{2+}-independent oxidation of simple amines and

phenolic monomers (Perez-Boada et al., 2005). The catalytic versatility of VP permits its application in Mn^{3+}-mediated or Mn–independent reactions on both low and high–redox potential aromatic substrates. Although VP from *P. eryngii* catalyzes the oxidation of Mn^{2+} to Mn^{3+} by H_2O_2, it differs from classical MnPs by its manganese-independent activity, enabling it to oxidize substituted phenols and synthetic dyes, as well as the LiP substrate, VA (Camarero et al., 1999).

3.2.2. DyP-type Peroxidases

The DyP (EC 1.11.1.19) family constitutes a new class of heme peroxidases and is found in fungi and bacteria (Sugano, 2009). DyPs show no similarity in primary sequence or structure to the other plant, bacterial or fungal peroxidases. DyP proteins have wide substrate specificity and function well under much lower pH conditions compared with plant peroxidases. They oxidize the typical peroxidase substrates, in addition to synthetic high–redox potential dyes of the anthraquinone type, which are not converted by the other peroxidases. DyP from the jelly fungus, *Auricularia auricula-judae*, was shown to oxidize VA and the non-phenolic β–O–4 dimeric lignin model compound at low pH, indicating the lignin-degrading activity of DyP. Although its ligninolytic activity has been described, the actual physiological role of DyPs is still unclear, and further studies are needed for their production on natural lignocellulose (Sugano, 2009).

3.2.3. Laccases

Laccase (benzenediol: oxygen oxidoreductase, EC 1.10.3.2) belongs to the superfamily of multi-copper oxidases and is found in plants, fungi, and bacteria (Dwivedi, Singh, Pandey, & Kumar, 2011). Plant laccase was identified in a wide variety of plants and is involved in wound response and the polymerization of lignin (Gavnholt & Larsen, 2002). Laccase is widely distributed in the wood-degrading fungi such as *T. versicolor, Trametes hirsuta, Trametes ochracea, Trametes villosa, Trametes gallica, Cerrena maxima, P. radiata, C. subvermispora*, and *P. eryngii* (Baldrian, 2006). Although, the majority of laccases produced by wood-degrading fungi are extracellular glycoproteins, intracellular laccases have also been reported (Dittmer, Patel, Dhawale, & Dhawale, 1997). Fungal laccase is reported to play roles in several processes including lignin degradation, morphogenesis, and pathogenesis (Thurston, 1994). The first bacterial laccase was found in the plant-root-associated bacterium *Azospirillum lipoferum* (Faure, Bouillant, & Bally, 1995). Laccase activity has also been demonstrated in other bacteria such as *Bacillus subtilis* (Martins et al., 2002), *Streptomyces cyaneus* (Arias et al., 2003), *Streptomyces lavendulae* (Suzuki et al., 2003) *Streptomyces griseus* (Endo et al., 2003) *Streptomyces coelicolor* (Machczynski, Vijgenboom, Samyn, & Canters, 2004), and *Thermus thermophilus* (Liu et al., 2011). Most of the studied bacterial laccases are intracellularly localized, such as in the case of *A. lipoferum* and *B. subtilis*, but extracellular bacterial laccases were also found as in the case of *S. cyaneus* (Arias et al., 2003). All fungal laccases show a similar structure consisting of three sequentially arranged copper domains. In addition to the regular three-domain laccases, small two-domain laccases have been recently described in bacteria (Ausec, van Elsas, & Mandic-Mulec, 2011; Skalova et al., 2009).

Laccases catalyze a one-electron oxidation with the concomitant four-electron reduction of molecular oxygen to water. Laccase catalysis requires four copper atoms arranged in two sites: blue type-1 (T1) mononuclear copper center and a trinuclear copper cluster consisting of one type-2 (T2) or normal copper center and two type-3 (T3) or coupled binuclear copper center. The four copper atoms are held in place in the catalytic center of the enzyme by four histidine-rich copper-binding regions (Claus, 2004). The different copper atoms differ in their spectroscopic and paramagnetic properties. The T1 copper is characterized by a strong absorption around 600 nm giving the enzyme its blue color, whereas the T2 copper exhibits only weak absorption in the visible region. The mononuclear T1 copper is the primary electron acceptor site where the enzyme catalyzes four 1-electron oxidations of the substrate. The electrons extracted from the reducing substrate are transferred to the T2/T3 trinuclear center where oxygen is reduced to water.

Fungal laccases (especially white-rot fungi) have higher redox potential compared with bacterial laccases, which are characterized by low-redox potential (Frasconi, Favero, Boer, Koivula, & Mazzei, 2010). Several lignin-degrading white-rot fungi, such as *P. chrysosporium*, *T. versicolor*, and *P. radiata*, produce laccase in addition to either LiP and/or MnP. However, the white-rot fungus *Pycnoporus cinnabarinus* lacks both LiP and MnP but produces laccase as the predominant enzyme and still degrades lignin efficiently (Eggert, Temp, & Eriksson, 1997).

3.2.3.1. Laccase/mediator System

Fungal laccases have been shown to degrade phenolic lignin model compounds (Kawai, Umezawa, & Higuchi, 1988). On the other hand, laccases can oxidize non-phenolic lignin model compounds only in the presence of redox mediators (Barreca, Fabbrini, Galli, Gentili, & Ljunggren, 2003; Li, Xu, & Eriksson, 1999). The degradation of β-O-4 non-phenolic lignin model compounds have been studied in the presence of synthetic redox mediators including 2,2′-azinobis-(3-ethylbenzthiazoline-6-sulfonate) (ABTS), 1-hydroxybenzotriazole, and violuric acid (Bourbonnais & Paice, 1992; Kawai, Nakagawa, & Ohashi, 2002; Li et al., 1999).

The efficiency of laccase-mediator systems to degrade recalcitrant compounds has a setback of high cost and possible toxicity of the artificial mediators, and this limits their application on an industrial scale. The potential of several natural phenolic compounds, related to the lignin polymer, to mediate the oxidative reactions catalyzed by laccase, has been tested with results showing that they efficiently promote *in vitro* oxidation of recalcitrant aromatic compounds (Camarero, Ibarra, Martinez, & Martinez, 2005).

3.2.4. Accessory Enzymes

White-rot fungi require sources of extracellular H_2O_2 to support the oxidative turnover of the LiPs and MnPs responsible for lignin degradation (Hammel, 1997). Thus, these fungi produce oxidases that generate the H_2O_2 required by their peroxidases (Fig. 1.7). Examples include the Aryl-alcohol oxidase found in *P. eryngii* (Guillen, Martinez, & Martinez, 1992) and the glyoxal oxidase found in *P. chrysosporium* (Kersten & Kirk 1987). In addition, fungi produce dehydrogenases that reduce lignin-derived compounds such as aryl-alcohol dehydrogenases and quinone reductases (Guillen, Martinez, Munoz, & Martinez, 1997). Moreover, many different fungi produce cellobiose dehydrogenase, an enzyme that degrades lignin in the presence of H_2O_2 and chelated Fe ions, generating hydroxyl radicals by Fenton-type reactions (Henriksson, Johansson, & Pettersson, 2000).

3.3. Lignin Degradation by Bacteria

Unlike the extensive studies on the microbial degradation of lignin by white-rot and brown-rot fungi, fewer reports are available on the depolymerization of lignin by soil bacteria (Bugg, Ahmad, Hardiman, & Rahmanpour, 2011; Crawford, Pometto, & Crawford, 1983; Kirby, 2006, pp. 125–168; Mercer, Iqbal, Miller, & Mccarthy, 1996).

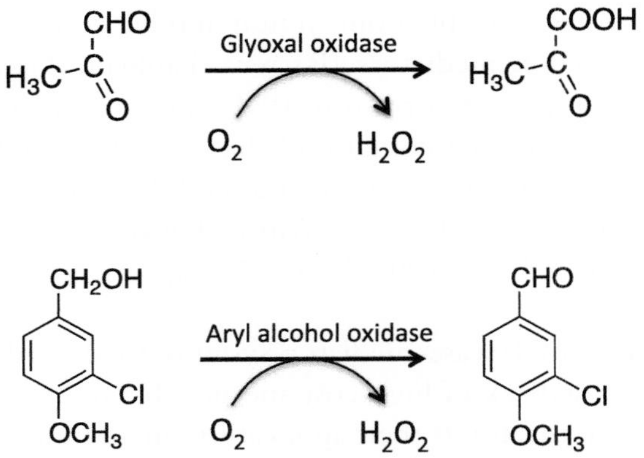

Figure 1.7 Reactions by peroxide-producing enzymes.

3.3.1. *Bacterial Peroxidases*

Streptomyces viridosporus T7A depolymerizes lignin producing a modified water-soluble, acid precipitable polymeric lignin (APPL) and releasing several single-ring aromatic phenols (Crawford, 1978; Crawford et al., 1983). Rüttimann, Seelenfreund, and Vicuña (1987) have shown that *S. viridosporus* was able to modify single-ring aromatic substrates but could not detect the cleavage of dimeric lignin model compounds. An extracellular LiP–designated ALiP-P3 that was able to oxidize phenolic and non-phenolic lignin model compounds was characterized in the supernatant of *S. viridosporus* T7A growing on lignocellulose (Ramachandra, Crawford, & Pometto, 1987). However, Spiker, Crawford, and Thiel (1992) showed that *S. viridosporus* LiP was capable of oxidizing phenolic but unable to oxidize non-phenolic compounds.

Wang, Bleakley, Crawford, Hertel, and Rafii (1990) reported the cloning of a 4.1 kb fragment from *S. viridosporus* genomic DNA expressing LiP activity in *Streptomyces lividans*. When the 4.1 kb fragment containing the LiP gene was cloned into *Pichia pastoris*, two activities of extracellular peroxidase and extracellular endoglucanase were detected, which suggested that the genes for lignocellulose degradation may be in a cluster on the *S. viridosporus* genome (Thomas & Crawford 1998). However, the nucleotide sequence of this fragment has not been published and hence is not available for comparative studies (Kirby, 2006, pp. 125–168).

Crude peroxidase preparations from *S. viridosporus* T7A was shown to have decolorizing activity against several azo dye isomers similar to the activity of fungal Mn-peroxidases. These results suggested that *S. viridosporus* T7A may produce an Mn-peroxidase or other peroxidase with similar substrate specificity to Mn-peroxidase (Burke & Crawford 1998). *Streptomyces badius* grown on lignocellulose was shown to generate APPL over a shorter period of time and with greater yields than that produced by *S. viridosporus*. *Streptomyces badius* peroxidase reacted with the *S. viridosporus* ALip-P3 antibody (Magnuson, Roberts, Crawford, & Hertel, 1991). However, no peroxidase has been purified or characterized from *S. badius*. Zimmerman, Umezawa, Broda, and Higuchi (1988) described the degradation of a non-phenolic β-O-4 lignin substructure model compound by *S. cyaneus*. An intracellular catalase–peroxidase was purified from this bacterium; however, the enzyme did not cleave the lignin model compounds (Mliki & Zimmermann, 1992).

Streptomyces albus ATCC 3005 was reported to produce higher levels of extracellular peroxidase activity than other actinomycetes (Antonopoulos, Rob, et al., 2001). The peroxidase was able to catalyze a broad range of substrates including 2,4-DCP, L-3,4-dihydroxyphenylalanine, 2,4,5-trichlorophenol, and other chlorophenols in the presence of H_2O_2 (Antonopoulos, Hernandez, et al., 2001). The enzyme was not purified, and the activity was not tested against lignin model compounds.

Two heme-proteins present in the supernatants of cultures of *Amycolatopsis* sp. 75iv2 ATCC 39,116 (formerly *Streptomyces setonii* and *S. griseus* 75vi2) grown in the presence of lignocellulose were characterized (Brown, Walker, Nakashige, Iavarone, & Chang, 2011). The enzymes were a catalase–peroxidase (Amyco1) and a catalase (Amyco2). Amyco1 was shown to degrade a phenolic but not a non-phenolic dimeric lignin model compound in the presence of ABTS and H_2O_2 (Brown et al., 2011).

Thermobifida fusca was also reported to partially degrade lignin when incubated with lignocellulosic pulps (Crawford, 1974). A DyP-type peroxidase was identified in the secretome of *T. fusca* grown in the presence of lignin (Adav, Ng, Arulmani, & Sze, 2010). A monomeric, heme-containing, thermostable and tat-dependently exported Dyp-type peroxidase (TfuDyp) was characterized. TfuDyp showed activity on guaiacol, 2,6-dimethoxyphenol (DMP), veratryl alcohol, *o*-phenylenediamine, and 3,3-diaminobenzidine (Van Bloois, Torres Pazmiño, Winter, & Fraaije, 2010).

Two dyp-peroxidases, DypA and DypB from *Rhodococcus jostii* RHA1, were purified and characterized (Ahmad et al., 2011). Using a nitrated lignin UV–vis assay, only DypB was activated by Mn^{2+} and exhibited activity in the presence of H_2O_2 (Ahmad et al., 2010). DypB was also reported to cleave Cα–Cβ bond in a lignin model compound and oxidize polymeric lignin (Ahmad et al., 2011).

3.3.2. Bacterial Laccases

A laccase was isolated, purified, and characterized from *S. cyaneus* (Arias et al., 2003). The oxidation of VA by this enzyme in the presence of ABTS as mediator was the first description of oxidation of non-phenolic substrate by a purified bacterial laccase. An extracellular laccase (SilA) produced by *Streptomyces ipomoea* CECT 3341 was cloned, overexpressed, and characterized. The enzyme decolorized and detoxified an azo–type dye and showed activity on a wide range of phenolic compounds including the syringyl and guaiacyl moieties derived from lignin. The coordinated action of SilA and acetosyringone (as mediator) resulted in the complete detoxification of the azo–type dye Orange II (Molina–Guijarro et al., 2009).

A two-domain extracellular laccase produced by *S. coelicolor* with activity against DMP at alkaline pH has been characterized. The gene encoding the *S. coelicolor* laccase was cloned, overexpressed, and purified (Machczynski et al., 2004). The enzyme was able to decolorize indigo carmine in the presence of syringaldehyde as a redox mediator (Dube, Shareck, Hurtubise, Beauregard, et al., 2008, Dube, Shareck, Hurtubise, Daneault, et al., 2008). When acetosyringone, a natural redox mediator, was used, the *S. coelicolor* laccase decolorized various dyes including acid blue 74, direct sky blue 6b, and reactive black 5 (Dube, Shareck, Hurtubise, Beauregard, et al., 2008, Dube, Shareck, Hurtubise, Daneault, et al., 2008).

The recombinant Tth-laccase from the thermophilic aerobic bacterium, *T. thermophilus* HB27, has been overexpressed in *Escherichia coli*. The Tth-laccase demonstrated pulp biobleaching activity in the presence of redox mediators. The activity was comparable to that described for bacterial laccases from *S. cyaneus* CECT 3335 (Zheng, Li, Li, & Shao, 2012). Similarly, the laccase from Gamma-proteobacterium JB, in the presence of ABTS, enhanced the brightness and reduced the kappa number of wheat straw-rich soda pulp (Singh, Ahuja, Batish, Capalash, & Sharma, 2008).

The laccase from *Bacillus licheniformis* was reported to oxidize ABTS syringaldazine and 2,6-DMP; however, the enzyme did not oxidize coumaric acid, cinnamic acid, and vanillic acid. Interestingly, the *B. lichenifor-mis* enzyme dimerized sinapic acid, caffeic acid, and ferulic acid (Koschor-reck et al., 2008).

4. BIOTECHNOLOGICAL APPLICATIONS OF LIGNINOLYTIC ENZYMES

Lignin-degrading enzymes have significant potential in industrial and biotechnological applications. The use of lignin-degrading enzymes in the pretreatment of recalcitrant lignocellulosic biomass would provide an environmentally friendly alternative to biofuel production compared to the thermal and chemical pretreatment techniques for the biofuel production (Wan & Li, 2012; Weng, Li, Bonawitz, & Chapple 2008). LiP, MnP, VP, and laccase can be used in the delignification and bioleaching of wood pulp to replace the traditional non-environmentally friendly chlorine-based delignification. They can also be applied in the decolorization of the dye wastewater from the textile industry. DyP peroxidases are espe-cially useful because of their ability to degrade anthraquinone dyes that are widely used by the textile industry (Sugano, 2009). Research on the biotechnological applications of laccase and laccase/mediator system has attracted much attention recently due to their ecofriendly nature as they use oxygen as electron donor and produce water as the only reaction byproduct (Riva, 2006). They have wide applications in the paper, textile, and food industries (Canas & Camarero, 2010). We discuss some of these applications below.

4.1. Delignification of Lignocellulose

Production of ethanol as alternative fuel using lignocellulosic substrates as raw materials is one of the most desirable goals to overcome the fossil fuel crisis. The transformation of lignocellulose into ethanol is achieved in three steps: (a) delignification to release cellulose and hemicellulose from their complex with lignin, (b) depolymerization of the carbohydrate polymers to produce free sugars, and (c) fermentation to ethanol using the liberated sugars. Biological treatment using white-rot fungi or other ligninolytic microorganisms including *Streptomyces* has been proposed, to replace the physicochemical treatments. Biological treatment can also be used for the removal of inhibitors prior to the fermentation. The advan-tages of using biological treatment include (i) mild reaction conditions,

(ii) higher product yields, (iii) fewer side reactions, and (iv) less energy demand (Lee, 1997).

4.2. Biopulping and Biobleaching

Lignin removal is important in the pulping and paper industry. Biopulping is the treatment of wood chips with lignin-degrading microorganisms to alter the lignin in the cell walls of wood, making the wood chips softer. This treatment not only improves paper strength and remove wood extractives but also reduces the energy consumption in the process of pulping. The production of pulp uses mechanical or chemical processes or a combination of the two processes. Pretreatment of wood chips for mechanical and chemical pulping with white-rot fungi has been developed (Mendonça, Jara, González, Elissetche, & Freer, 2008). Laccases from white-rot fungi can be applied in biopulping to partially degrade the lignin and therefore loosen lignin structures (Mendonça et al., 2008). *Ceriporiopsis subvermispora* and *Pleurotus* are fungi used in biopulping (Pérez, Muñoz-Dorado, De La Rubia, & Martínez, 2002).

Biobleaching is the bleaching of pulps using enzymes or ligninolytic fungi that reduce the amount of chemical bleach required to obtain a desirable brightness of pulps. Laccase-mediator system has been shown to possess the potential to substitute for chlorine-containing reagents. Laccases can also be applied as biobleaching agents as they degrade lignin and decolorize the pulp (Call & Call, 2005). Laccase produced by *T. versicolor* has been studied for biobleaching of paper pulp and other industrial applications (Wesenberg, Kyriakides, Agathos, 2003). The role of lignin-degrading enzymes from *Streptomyces* in biopulping and biobleaching has also been studied. Biobleaching of eucalyptus Kraft pulp with *S. albus* culture supernatant in the presence of H_2O_2 resulted in a significant reduction of kappa number with no change in viscosity suggesting a potential application of *S. albus* in biobleaching (Antonopoulos, Hernandez, et al., 2001). *Streptomyces cyaneus* laccase was able to delignify Kraft pulp with ABTS as a mediator, indicating the potential application of the laccases from Streptomycetes in biobleaching of Kraft lignin in the presence of synthetic mediators (Arias et al., 2003). Biobleaching experiments carried out on *Eucalyptus globulus* Kraft pulps with *S. ipomoea* laccase in the presence of acetosyringone as a natural mediator also showed reduction in kappa number and increase of brightness without decreasing the viscosity values significantly (Eugenio et al., 2011). These results suggest significant promise for the use of Streptomycetes lignin-degrading enzymes in industrial application.

4.3. Textile Dye Transformation

The textile industry uses water as a medium for removing impurities, application of dyes and finishing agents. There is a significant water pollution associated with these processes due to the highly toxic dyes, bleaching agents, salt acids, and the alkali employed. LiP and MnP from the white-rot fungus *P. chrysosporium* have been investigated for their dye decolorization with the results showing the capacity to mineralize a variety of recalcitrant aromatic pollutants (Mehta, 2012). Decolorization of 23 industrial dyes by 16 white-rot fungi has also been investigated. The crude extracts of the cultures showed laccase, LiP, and aryl alcohol oxidase activities. However, only laccase activity was correlated with color removal (Rodríguez, Pickard, & Vazquez-Duhalt, 1999). Although some dyes are not degradable by laccases, many are oxidized by the enzyme and therefore initiating the destruction of the dyes (Schliephake, Mainwaring, Lonergan, Jones, & Baker, 2000).

An extensive review on the role of peroxidases in the treatment and decolorization of wide spectrum of aromatic dyes from polluted water can be found in the literature (Husain, 2010). Kirby, Marchant, and Mcmullan (2000) reported that laccase from *P. tremellosa* decolorized synthetic textile dyes. Also, laccases used in combination with mediators and cellobiose dehydrogenase were shown to be an ecofriendly alternative for chemical treatment of textile dye wastes (Ciullini, Tilli, Scozzafava, & Briganti, 2008).

4.4. Decolorization of Distillery Effluent and Waste Effluent Treatment

The characteristic of the dark brown appearance of distillery wastewater is mainly due to the high molecular weight organic compounds called melanoidin, a product of the Maillard reaction of sugars with proteins. The brown color is also due to the presence of phenolics from the feedstock, caramels from overheated sugars, and furfurals from acid hydrolysis (Kort, 1979, pp. 97–130). The detoxification and decolorization of this industrial waste is performed using oxidative enzymes (laccases and peroxidases) from bacteria, fungi, and yeast (Rajasundari & Murugesan, 2011). *Coriolus versicolor* was the first fungal strain shown to decolorize this type of waste (Watanabe, Sugi, & Tanaka, 1982). *P. chrisosporium* JAG-40 decolorized synthetic and natural melanoidin (Dahiya, Singh, & Nigam, 2001). Part of the treatment of these wastes includes the use of laccases and peroxidases that oxidize phenolic compounds to aryl-oxy radicals creating complexes that are insoluble. Other mechanisms carried out by these enzymes include the polymerization of the contaminants themselves or the copolymerization with other non-toxic substrates to facilitate their removal by sedimentation, adsorption, or filtration (Gianfreda, Iamarino, Scelza, & Rao, 2006; Rabinovich, Bolobova, & Vasil'chenko, 2004).

4.5. Degradation of Herbicides

The capacity of lignin-degrading enzymes to degrade herbicides has also been investigated. LiP and MnP produced by the white-rot fungus *P. chrysosporium* degraded the herbicide, isoproturon in *in vivo* and *in vitro* experiments (Del Pilar Castillo, Von Wirén-Lehr, Scheunert, & Torstensson, 2001). MnP from *P. chrysosporium* was able to oxidize bentazon in the presence of Mn(II) and Tween 80 (Castillo, Ander, Stenström, & Torstensson, 2000). The herbicide glyphosate was degraded by *Nematoloma frowardii* MnP and *T. versicolor* laccase in the presence of ABTS as a mediator (Pizzul, Castillo, & Stenström, 2009). These reports clearly show the potential application of lignin-degrading enzymes in the treatment of soil and drain water contaminated with herbicides.

5. CONCLUSION

In the foregoing discussion, we have provided insights into the current state of knowledge on lignin-degrading enzymes from both eukaryotic (mostly fungi) and bacterial sources. While there is a better understanding of the lignin-degrading processes in the eukaryotic organisms, the knowledge on bacterial lignin degradation is somewhat lagging. Recent interest in biofuel production has, however, re-energized research in enzymatic deconstruction of lignin, as it is seen to be more environmentally friendly in the production of next generation biofuels. We anticipate that the increased interest, with its concomitant increase in research funding will provide the catalyst for rapid and better understanding of this field that evolves so slowly. The new knowledge acquired should help open new areas of application of the fascinating processes evolved by nature to dismantle this highly recalcitrant biopolymer.

ACKNOWLEDGMENTS

AMA and JOS would like to thank Dr John Gerlt and Dr John Cronan and the Energy Biosciences Institute for supporting their research on lignin deconstruction.

REFERENCES

Adav, S. S., Ng, C. S., Arulmani, M., & Sze, S. K. (2010). Quantitative iTRAQ secretome analysis of cellulolytic *Thermobifida fusca*. *Journal of Proteome Research*, *9*(6), 3016–3024.

Ahmad, M., Roberts, J. N., Hardiman, E. M., Singh, R., Eltis, L. D., & Bugg, T. D. (2011). Identification of DypB from *Rhodococcus jostii* RHA1 as a lignin peroxidase. *Biochemistry*, *50*(23), 5096–5107.

Ahmad, M., Taylor, C. R., Pink, D., Burton, K., Eastwood, D., Bending, G. D., et al. (2010). Development of novel assays for lignin degradation: comparative analysis of bacterial and fungal lignin degraders. *Molecular BioSystems*, *6*(5), 815–821.

Amen-Chen, C., Pakdel, H., & Roy, C. (2001). Production of monomeric phenols by thermochemical conversion of biomass: a review. *Bioresource Technology*, *79*(3), 277–299.

Ander, P., & Marzullo, L. (1997). Sugar oxidoreductases and veratryl alcohol oxidase as related to lignin degradation. *Journal of Biotechnology*, *53*(2–3), 115–131.

Antonopoulos, V. T., Hernandez, M., Arias, M. E., Mavrakos, E., & Ball, A. S. (2001). The use of extracellular enzymes from *Streptomyces albus* ATCC 3005 for the bleaching of eucalyptus kraft pulp. *Applied Microbiology and Biotechnology*, *57*(1–2), 92–97.

Antonopoulos, V. T., Rob, A., Ball, A. S., & Wilson, M. T. (2001). Dechlorination of chlorophenols using extracellular peroxidases produced by *Streptomyces albus* ATCC 3005. *Enzyme and Microbial Technology*, *29*(1), 62–69.

Arias, M. E., Arenas, M., Rodríguez, J., Soliveri, J., Ball, A. S., & Hernández, M. (2003). Kraft pulp biobleaching and mediated oxidation of a nonphenolic substrate by laccase from *Streptomyces cyaneus* CECT 3335. *Applied and Environmental Microbiology*, *69*(4), 1953–1958.

Ausec, L., van Elsas, J. D., & Mandic-Mulec, I. (2011). Two- and three-domain bacterial laccase-like genes are present in drained peat soils. *Soil Biology and Biochemistry*, *43*(5), 975–983.

Baldrian, P. (2006). Fungal laccases-occurrence and properties. *FEMS Microbiology Reviews*, *30*(2), 215–242.

Bao, W., Fukushima, Y., Jensen, K. A., Jr., Moen, M. A., & Hammel, K. E. (1994). Oxidative degradation of non-phenolic lignin during lipid peroxidation by fungal manganese peroxidase. *FEBS Letters*, *354*(3), 297–300.

Barreca, A. M., Fabbrini, M., Galli, C., Gentili, P., & Ljunggren, S. (2003). Laccase/mediated oxidation of a lignin model for improved delignification procedures. *Journal of Molecular Catalysis B: Enzymatic*, *26*(1–2), 105–110.

Bietti, M., Baciocchi, E., & Steenken, S. (1998). Lifetime, reduction potential and base-induced fragmentation of the veratryl alcohol radical cation in aqueous solution. Pulse radiolysis studies on a ligninase "mediator". *Journal of Physical Chemistry A*, *102*(38), 7337–7342.

Boerjan, W., Ralph, J., & Baucher, M. (2003). Lignin biosynthesis. *Annual Review of Plant Biology*, *54*, 519–546.

Bourbonnais, R., & Paice, M. G. (1992). Demethylation and delignification of kraft pulp by *Trametes versicolor* laccase in the presence of 2,2'-azinobis-(3-ethylbenzthiazoline-6-sulphonate). *Applied Microbiology and Biotechnology*, *36*(6), 823–827.

Boyle, C. D., Kropp, B. R., & Reid, I. D. (1992). Solubilization and mineralization of lignin by white rot fungi. *Applied and Environmental Microbiology*, *58*(10), 3217–3224.

Brown, M. E., Walker, M. C., Nakashige, T. G., Iavarone, A. T., & Chang, M. C. (2011). Discovery and characterization of heme enzymes from unsequenced bacteria: application to microbial lignin degradation. *Journal of the American Chemical Society*, *133*(45), 18006–18009.

Bugg, T. D. H., Ahmad, M., Hardiman, E. M., & Rahmanpour, R. (2011). Pathways for degradation of lignin in bacteria and fungi. *Natural Product Reports*, *28*(12), 1883–1896.

Burke, N. S., & Crawford, D. L. (1998). Use of azo dye ligand chromatography for the partial purification of a novel extracellular peroxidase from *Streptomyces viridosporus* T7A. *Applied Microbiology and Biotechnology*, *49*(5), 523–530.

Call, H.-P., & Call, S. (2005). New generation of enzymatic delignification and bleaching. *Pulp and Paper Canada*, *106*(1), 45–48.

Camarero, S., Ibarra, D., Martinez, M. J., & Martinez, A. T. (2005). Lignin-derived compounds as efficient laccase mediators for decolorization of different types of recalcitrant dyes. *Applied and Environmental Microbiology*, *71*(4), 1775–1784.

Camarero, S., Sarkar, S., Ruiz-Dueñas, F. J., Martínez, M. J., & Martínez, A. T. (1999). Description of a versatile peroxidase involved in the natural degradation of lignin that has both manganese peroxidase and lignin peroxidase substrate interaction sites. *Journal of Biological Chemistry*, *274*(15), 10324–10330.

Canas, A. I., & Camarero, S. (2010). Laccases and their natural mediators: biotechnological tools for sustainable eco-friendly processes. *Biotechnology Advances*, *28*(6), 694–705.

Candeias, L. P., & Harvey, P. J. (1995). Lifetime and reactivity of the veratryl alcohol radical cation. Implications for lignin peroxidase catalysis. *The Journal of Biological Chemistry*, *270*(28), 16745–16748.

Castillo, M. D. P., Ander, P., Stenström, J., & Torstensson, L. (2000). Degradation of the herbicide bentazon as related to enzyme production by *Phanerochaete chrysosporium* in two solid substrate fermentation systems. *World Journal of Microbiology and Biotechnology*, *16*(3), 289–295.

Choinowski, T., Blodig, W., Winterhalter, K. H., & Piontek, K. (1999). The crystal structure of lignin peroxidase at 1.70 Å resolution reveals a hydroxy group on the Cβ of tryptophan 171: a novel radical site formed during the redox cycle. *Journal of Molecular Biology*, *286*(3), 809–827.

Ciullini, I., Tilli, S., Scozzafava, A., & Briganti, F. (2008). Fungal laccase, cellobiose dehydrogenase, and chemical mediators: combined actions for the decolorization of different classes of textile dyes. *Bioresource Technology*, *99*(15), 7003–7010.

Claus, H. (2004). Laccases: structure, reactions, distribution. *Micron*, *35*(1–2), 93–96.

Crawford, D. L. (1974). Growth of *Thermomonospora fusca* on lignocellulosic pulps of varying lignin content. *Canadian Journal of Microbiology*, *20*(7), 1069–1072.

Crawford, D. L. (1978). Lignocellulose decomposition by selected *Streptomyces* strains. *Applied and Environmental Microbiology*, *35*(6), 1041–1045.

Crawford, D. L., Pometto, A. L., III, & Crawford, R. L. (1983). Lignin degradation by *Streptomyces viridosporus*: isolation and characterization of a new polymeric lignin degradation intermediate. *Applied and Environmental Microbiology*, *45*(3), 898–904.

Dahiya, J., Singh, D., & Nigam, P. (2001). Decolourisation of synthetic and spentwash melanoidins using the white-rot fungus *Phanerochaete chrysosporium* JAG-40. *Bioresource Technology*, *78*(1), 95–98.

Del Pilar Castillo, M., Von Wirén-Lehr, S., Scheunert, I., & Torstensson, L. (2001). Degradation of isoproturon by the white rot fungus *Phanerochaete chrysosporium*. *Biology and Fertility of Soils*, *33*(6), 521–528.

Dittmer, J. K., Patel, N. J., Dhawale, S. W., & Dhawale, S. S. (1997). Production of multiple laccase isoforms by *Phanerochaete chryosporium* grown under nutrient sufficiency. *FEMS Microbiology Letters*, *149*(1), 65–70.

Dorrestijn, E., Laarhoven, L. J. J., Arends, I. W. C. E., & Mulder, P. (2000). The occurrence and reactivity of phenoxyl linkages in lignin and low rank coal. *Journal of Analytical and Applied Pyrolysis*, *54*(1–2), 153–192.

Doyle, W. A., Blodig, W., Veitch, N. C., Piontek, K., & Smith, A. T. (1998). Two substrate interaction sites in lignin peroxidase revealed by site-directed mutagenesis. *Biochemistry*, *37*(43), 15097–15105.

Dube, E., Shareck, F., Hurtubise, Y., Beauregard, M., & Daneault, C. (2008). Decolourization of recalcitrant dyes with a laccase from *Streptomyces coelicolor* under alkaline conditions. *Journal of Industrial Microbiology & Biotechnology*, *35*(10), 1123–1129.

Dube, E., Shareck, F., Hurtubise, Y., Daneault, C., & Beauregard, M. (2008). Homologous cloning, expression, and characterisation of a laccase from *Streptomyces coelicolor* and enzymatic decolourisation of an indigo dye. *Applied Microbiology and Biotechnology, 79*(4), 597–603.

Dwivedi, U. N., Singh, P., Pandey, V. P., & Kumar, A. (2011). Structure–function relationship among bacterial, fungal and plant laccases. *Journal of Molecular Catalysis B: Enzymatic, 68*(2), 117–128.

Eggert, C., Temp, U., & Eriksson, K.-.L. (1997). Laccase is essential for lignin degradation by the white-rot fungus *Pycnoporus cinnabarinus*. *FEBS Letters, 407*(1), 89–92.

Endo, K., Hayashi, Y., Hibi, T., Hosono, K., Beppu, T., & Ueda, K. (2003). Enzymological characterization of EpoA, a laccase-like phenol oxidase produced by *Streptomyces griseus*. *Journal of Biochemistry, 133*(5), 671–677.

Eugenio, M. E., Hernández, M., Moya, R., Martín-Sampedro, R., Villar, J. C., & Arias, M. E. (2011). Evaluation of a new laccase produced by *Streptomyces ipomoea* on biobleaching and ageing of kraft pulps. *BioResources, 6*(3), 3231–3241.

Faix, O. (1991). Classification of lignins from different botanical origins by FT-IR spectroscopy. *Holzforschung, 45*, 21–27.

Faure, D., Bouillant, M., & Bally, R. (1995). Comparative study of substrates and inhibitors of *Azospirillum lipoferum* and *Pyricularia oryzae* laccases. *Applied and Environmental Microbiology, 61*(3), 1144–1146.

Foyle, T., Jennings, L., & Mulcahy, P. (2007). Compositional analysis of lignocellulosic materials: evaluation of methods used for sugar analysis of waste paper and straw. *Bioresource Technology, 98*(16), 3026–3036.

Frasconi, M., Favero, G., Boer, H., Koivula, A., & Mazzei, F. (2010). Kinetic and biochemical properties of high and low redox potential laccases from fungal and plant origin. *Biochimica et Biophysica Acta (BBA) – Proteins & Proteomics, 1804*(4), 899–908.

Gavnholt, B., & Larsen, K. (2002). Molecular biology of plant laccases in relation to lignin formation. *Physiologia Plantarum, 116*(3), 273–280.

Gianfreda, L., Iamarino, G., Scelza, R., & Rao, M. A. (2006). Oxidative catalysts for the transformation of phenolic pollutants: a brief review. *Biocatalysis and Biotransformation, 24*(3), 177–187.

Gilbertson, R. L. (1980). Wood-rotting fungi of North America. *Mycologia, 72*(1), 1–49.

Glenn, J. K., & Gold, M. H. (1985). Purification and characterization of an extracellular Mn(II)-dependent peroxidase from the lignin-degrading basidiomycete, *Phanerochaete chrysosporium*. *Archives of Biochemistry and Biophysics, 242*(2), 329–341.

Guillen, F., Martinez, A. T., & Martinez, M. J. (1992). Substrate specificity and properties of the aryl-alcohol oxidase from the ligninolytic fungus *Pleurotus eryngii*. *European Journal of Biochemistry/FEBS, 209*(2), 603–611.

Guillen, F., Martinez, M. J., Munoz, C., & Martinez, A. T. (1997). Quinone redox cycling in the ligninolytic fungus *Pleurotus eryngii* leading to extracellular production of superoxide anion radical. *Archives of Biochemistry and Biophysics, 339*(1), 190–199.

Hammel, K. E. (1997). Fungal degradation of ligni. In G. K. E. Cadisch & G. (Eds.), *Driven by nature: Plant litter quality and decomposition* (pp. 33–45). Oxon: CAB International.

Harris, D., & Debolt, S. (2010). Synthesis, regulation and utilization of lignocellulosic biomass. *Plant Biotechnology Journal, 8*(3), 244–262.

Harvey, P. J., Schoemaker, H. E., & Palmer, J. M. (1986). Veratryl alcohol as a mediator and the role of radical cations in lignin biodegradation by *Phanerochaete chrysosporium*. *FEBS Letters, 195*(1–2), 242–246.

Heinfling, A., Ruiz-Dueñas, F. J., Martínez, M. J., Bergbauer, M., Szewzyk, U., & Martínez, A. T. (1998). A study on reducing substrates of manganese-oxidizing peroxidases from *Pleurotus eryngii* and *Bjerkandera adusta*. *FEBS Letters, 428*(3), 141–146.

Henriksson, G., Johansson, G., & Pettersson, G. (2000). A critical review of cellobiose dehydrogenases. *Journal of Biotechnology, 78*(2), 93–113.

Hofrichter, M. (2002). Review: lignin conversion by manganese peroxidase (MnP). *Enzyme and Microbial Technology, 30*(4), 454–466.

Hofrichter, M., Ullrich, R., Pecyna, M. J., Liers, C., & Lundell, T. (2010). New and classic families of secreted fungal heme peroxidases. *Applied Microbiology and Biotechnology, 87*(3), 871–897.

Husain, Q. (2010). Peroxidase mediated decolorization and remediation of wastewater containing industrial dyes: a review. *Reviews in Environmental Science and Biotechnology, 9*(2), 117–140.

Johansson, T., Welinder, K. G., & Nyman, P. O. (1993). Isozymes of lignin peroxidase and manganese(II) peroxidase from the white-rot basidiomycete *Trametes versicolor*. II. Partial sequences, peptide maps, and amino acid and carbohydrate compositions. *Archives of Biochemistry and Biophysics, 300*(1), 57–62.

Kapich, A., Hofrichter, M., Vares, T., & Hatakka, A. (1999). Coupling of manganese peroxidase-mediated lipid peroxidation with destruction of nonphenolic lignin model compounds and14C-labeled lignins. *Biochemical and Biophysical Research Communications, 259*(1), 212–219.

Kawai, S., Nakagawa, M., & Ohashi, H. (2002). Degradation mechanisms of a nonphenolic β-O-4 lignin model dimer by *Trametes versicolor* laccase in the presence of 1-hydroxyben-zotriazole. *Enzyme and Microbial Technology, 30*(4), 482–489.

Kawai, S., Umezawa, T., & Higuchi, T. (1988). Degradation mechanisms of phenolic β-1 lignin substructure model compounds by laccase of *Coriolus versicolor*. *Archives of Biochemistry and Biophysics, 262*(1), 99–110.

Kersten, P. J., & Kirk, T. K. (1987). Involvement of a new enzyme, glyoxal oxidase, in extracellular H_2O_2 production by *Phanerochaete chrysosporium*. *Journal of Bacteriology, 169*(5), 2195–2201.

Khindaria, A., Yamazaki, I., & Aust, S. D. (1996). Stabilization of the veratryl alcohol cation radical by lignin peroxidase. *Biochemistry, 35*(20), 6418–6424.

Kirby, R. (2006). *Actinomycetes and lignin degradation. Advances in applied microbiology.* Academic Press.

Kirby, N., Marchant, R., & Mcmullan, G. (2000). Decolourisation of synthetic textile dyes by *Phlebia tremellosa. FEMS Microbiology Letters, 188*(1), 93–96.

Kirk, T. K., & Farrell, R. L. (1987). Enzymatic "combustion": the microbial degradation of lignin. *Annual Review of Microbiology, 41*, 465–505.

Kirk, T. K., Tien, M., & Kersten, P. J. (1986). Ligninase of *Phanerochaete chrysosporium*. Mechanism of its degradation of the non-phenolic arylglycerol β-aryl ether substructure of lignin. *Biochemical Journal, 236*(1), 279–287.

Kleinert, M., & Barth, T. (2008). Phenols from lignin. *Chemical Engineering and Technology, 31*(5), 736–745.

Kort, M. J. (1979). *Colour in the sugar industry. Science and Technology.* London: Applied Science.

Koschorreck, K., Richter, S. M., Ene, A. B., Roduner, E., Schmid, R. D., & Urlacher, V. B. (2008). Cloning and characterization of a new laccase from *Bacillus licheniformis* catalyzing dimerization of phenolic acids. *Applied Microbiology and Biotechnology, 79*(2), 217–224.

Lee, J. (1997). Biological conversion of lignocellulosic biomass to ethanol. *Journal of Biotechnology, 56*(1), 1–24.

Leonowicz, A., Cho, N. S., Luterek, J., Wilkolazka, A., Wojtas-Wasilewska, M., Matuszewska, A., et al. (2001). Fungal laccase: properties and activity on lignin. *Journal of Basic Microbiology, 41*(3–4), 185–227.

Leonowicz, A., Matuszewska, A., Luterek, J., Ziegenhagen, D., Wojtaś-Wasilewska, M., Cho, N., et al. (1999). Biodegradation of lignin by white rot fungi. *Fungal Genetics and Biology, 27*(2–3), 175–185.

Liers, C., Bobeth, C., Pecyna, M., Ullrich, R., & Hofrichter, M. (2010). DyP-like peroxidases of the jelly fungus *Auricularia auricula-judae* oxidize nonphenolic lignin model compounds and high-redox potential dyes. *Applied Microbiology and Biotechnology, 85*(6), 1869–1879.

Liu, X., Gillespie, M., Ozel, A. D., Dikici, E., Daunert, S., & Bachas, L. G. (2011). Electrochemical properties and temperature dependence of a recombinant laccase from *Thermus thermophilus*. *Analytical and Bioanalytical Chemistry, 399*(1), 361–366.

Li, K., Xu, F., & Eriksson, K. E. (1999). Comparison of fungal laccases and redox mediators in oxidation of a nonphenolic lignin model compound. *Applied and Environmental Microbiology, 65*(6), 2654–2660.

Lobos, S., Larrain, J., Salas, L., Cullen, D., & Vicuna, R. (1994). Isoenzymes of manganese-dependent peroxidase and laccase produced by the lignin-degrading basidiomycete *Ceriporiopsis subvermispora*. *Microbiology, 140*(10), 2691–2698.

Lundell, T., Wever, R., Floris, R., Harvey, P., Hatakka, A., Brunow, G., et al. (1993). Lignin peroxidase L3 from *Phlebia radiata*. Pre-steady-state and steady-state studies with veratryl alcohol and a non-phenolic lignin model compound 1-(3,4-dimethoxyphenyl)-2-(2-methoxyphenoxy)propane-1,3-diol. *European Journal of Biochemistry/FEBS, 211*(3), 391–402.

Machczynski, M. C., Vijgenboom, E., Samyn, B., & Canters, G. W. (2004). Characterization of SLAC: a small laccase from *Streptomyces coelicolor* with unprecedented activity. *Protein Science: A Publication of the Protein Society, 13*(9), 2388–2397.

Magnuson, T. S., Roberts, M. A., Crawford, D. L., & Hertel, G. (1991). Immunologic relatedness of extracellular ligninases from the actinomycetes *Streptomyces viridosporus* t7a and *Streptomyces badius* 252. *Applied Biochemistry and Biotechnology, 28–29*(1), 433–443.

Martinez, M. J., Böckle, B., Camarero, S., Guillén, F., & Martinez, A. T. (1996). *MnP isoenzymes produced by two Pleurotus species in liquid culture and during wheat-straw solid-state fermentation*.

Martinez, D., Larrondo, L. F., Putnam, N., Gelpke, M. D., Huang, K., Chapman, J., et al. (2004). Genome sequence of the lignocellulose degrading fungus *Phanerochaete chrysosporium* strain RP78. *Nature Biotechnology, 22*(6), 695–700.

Martinez, A. T., Speranza, M., Ruiz-Duenas, F. J., Ferreira, P., Camarero, S., Guillen, F., et al. (2005). Biodegradation of lignocellulosics: microbial, chemical, and enzymatic aspects of the fungal attack of lignin. *International Microbiology: The Official Journal of the Spanish Society for Microbiology, 8*(3), 195–204.

Martins, L. O., Soares, C. M., Pereira, M. M., Teixeira, M., Costa, T., Jones, G. H., et al. (2002). Molecular and biochemical characterization of a highly stable bacterial laccase that occurs as a structural component of the *Bacillus subtilis* endospore coat. *Journal of Biological Chemistry, 277*(21), 18849–18859.

Mehta, R. (2012). Bioremediation of textile waste water. *Colourage, 59*(4), 46.

Mendonça, R. T., Jara, J. F., González, V., Elissetche, J. P., & Freer, J. (2008). Evaluation of the white-rot fungi *Ganoderma australe* and *Ceriporiopsis subvermispora* in biotechnological applications. *Journal of Industrial Microbiology and Biotechnology, 35*(11), 1323–1330.

Mercer, D. K., Iqbal, M., Miller, P., & Mccarthy, A. J. (1996). Screening actinomycetes for extracellular peroxidase activity. *Applied and Environmental Microbiology, 62*(6), 2186–2190.

Miki, K., Renganathan, V., & Gold, M. H. (1986). Mechanism of β-aryl ether dimeric lignin model compound oxidation by lignin peroxidase of *Phanerochaete chrysosporium*. *Biochemistry, 25*(17), 4790–4796.

Millis, C. D., Cai, D., Stankovich, M. T., & Tien, M. (1989). Oxidation–reduction potentials and ionization states of extracellular peroxidases from the lignin-degrading fungus *Phanerochaete chrysosporium*. *Biochemistry, 28*(21), 8484–8489.

Mliki, A., & Zimmermann, W. (1992). Purification and characterization of an intracellular peroxidase from *Streptomyces cyaneus*. *Applied and Environmental Microbiology, 58*(3), 916–919.

Moilanen, A. M., Lundell, T., Vares, T., & Hatakka, A. (1996). Manganese and malonate are individual regulators for the production of lignin and manganese peroxidase isozymes and in the degradation of lignin by *Phlebia radiata*. *Applied Microbiology and Biotechnology, 45*(6), 792–799.

Molina-Guijarro, J. M., Pérez, J., Muñoz-Dorado, J., Guillén, F., Moya, R., Hernńdez, M., et al. (2009). Detoxification of azo dyes by a novel pH-versatile, salt-resistant laccase from *Streptomyces ipomoea*. *International Microbiology, 12*(1), 13–21.

Monrroy, M., Ortega, I., Ramírez, M., Baeza, J., & Freer, J. (2011). Structural change in wood by brown rot fungi and effect on enzymatic hydrolysis. *Enzyme and Microbial Technology, 49*(5), 472–477.

Moreira, P. R., Almeida-Vara, E., Malcata, F. X., & Duarte, J. C. (2007). Lignin transformation by a versatile peroxidase from a novel *Bjerkandera* sp. strain. *International Biodeterioration & Biodegradation, 59*(3), 234–238.

Pandey, M. P., & Kim, C. S. (2011). Lignin depolymerization and conversion: a review of thermochemical methods. *Chemical Engineering and Technology, 34*(1), 29–41.

Parthasarathi, R., Romero, R. A., Redondo, A., & Gnanakaran, S. (2011). Theoretical study of the remarkably diverse linkages in lignin. *Journal of Physical Chemistry Letters, 2*(20), 2660–2666.

Paszczyński, A., Huynh, V., & Crawford, R. (1985). Enzymatic activities of an extracellular, manganese-dependent peroxidase from *Phanerochaete chrysosporium*. *FEMS Microbiology Letters, 29*(1–2), 37–41.

Perez-Boada, M., Ruiz-Duenas, F. J., Pogni, R., Basosi, R., Choinowski, T., Martinez, M. J., et al. (2005). Versatile peroxidase oxidation of high redox potential aromatic compounds: site-directed mutagenesis, spectroscopic and crystallographic investigation of three long-range electron transfer pathways. *Journal of Molecular Biology, 354*(2), 385–402.

Pérez, J., Muñoz-Dorado, J., De La Rubia, T., & Martínez, J. (2002). Biodegradation and biological treatments of cellulose, hemicellulose and lignin: an overview. *International Microbiology, 5*(2), 53–63.

Piontek, K., Smith, A. T., & Blodig, W. (2001). Lignin peroxidase structure and function. *Biochemical Society Transactions, 29*(Pt 2), 111–116.

Pizzul, L., Castillo, M. D.P., & Stenström, J. (2009). Degradation of glyphosate and other pesticides by ligninolytic enzymes. *Biodegradation, 20*(6), 751–759.

Poulos, T. L., Edwards, S. L., Wariishi, H., & Gold, M. H. (1993). Crystallographic refinement of lignin peroxidase at 2 A. *The Journal of Biological Chemistry, 268*(6), 4429–4440.

Rabinovich, M. L., Bolobova, A. V., & Vasil'chenko, L. G. (2004). Fungal decomposition of natural aromatic structures and xenobiotics: a review. *Applied Biochemistry and Microbiology, 40*(1), 1–17.

Rajasundari, K., & Murugesan, R. (2011). Decolourization of distillery waste water – role of microbes and their potential oxidative enzymes (review). *Journal of Applied Environmental and Biological Sciences, 1*(4), 54–68.

Ramachandra, M., Crawford, D. L., & Pometto, A. L. (1987). Extracellular enzyme activities during lignocellulose degradation by streptomyces spp.: a comparative study of wild-type and genetically manipulated strains. *Applied and Environmental Microbiology, 53*(12), 2754–2760.

Reale, S., Di Tullio, A., Spreti, N., & De Angelis, F. (2004). Mass spectrometry in the biosynthetic and structural investigation of lignins. *Mass Spectrometry Reviews, 23*(2), 87–126.

Reddy, G.V.B., Sridhar, M., & Gold, M. H. (2003). Cleavage of nonphenolic β-1 diarylpropane lignin model dimers by manganese peroxidase from *Phanerochaete chrysosporium*: evidence for a hydrogen abstraction mechanism. *European Journal of Biochemistry, 270*(2), 284–292.

Riva, S. (2006). Laccases: blue enzymes for green chemistry. *Trends in Biotechnology, 24*(5), 219–226.

Rodríguez, E., Pickard, M. A., & Vazquez-Duhalt, R. (1999). Industrial dye decolorization by laccases from ligninolytic fungi. *Current Microbiology, 38*(1), 27–32.

Ruiz-Duenas, F. J., Morales, M., Garcia, E., Miki, Y., Martinez, M. J., & Martinez, A. T. (2009). Substrate oxidation sites in versatile peroxidase and other basidiomycete peroxidases. *Journal of Experimental Botany, 60*(2), 441–452.

Ruttimann-Johnson, C., Salas, L., Vicuna, R., & Kirk, T. K. (1993). Extracellular enzyme production and synthetic lignin mineralization by *Ceriporiopsis subvermispora*. *Applied and Environmental Microbiology, 59*(6), 1792–1797.

Rüttimann, C., Seelenfreund, D., & Vicuña, R. (1987). Metabolism of low molecular weight lignin-related compounds by *Streptomyces viridosporus* T7A. *Enzyme and Microbial Technology, 9*(9), 526–530.

Scheller, H. V., & Ulvskov, P. (2010). Hemicelluloses. *Annual Review of Plant Biology, 61,* 263–289.

Schliephake, K., Mainwaring, D. E., Lonergan, G. T., Jones, I. K., & Baker, W. L. (2000). Transformation and degradation of the disazo dye Chicago sky blue by a purified laccase from *Pycnoporus cinnabarinus*. *Enzyme and Microbial Technology, 27*(1–2), 100–107.

Schoemaker, H. E., Lundell, T., Hatakka, A., & Piontek, K. (1994). The oxidation of veratryl alcohol, dimeric lignin models and lignin by lignin peroxidase: the redox cycle revisioned. *FEMS Microbiology Reviews, 13,* 321–332.

Singh, G., Ahuja, N., Batish, M., Capalash, N., & Sharma, P. (2008). Biobleaching of wheat straw-rich soda pulp with alkalophilic laccase from gamma-proteobacterium JB: optimization of process parameters using response surface methodology. *Bioresource Technology, 99*(16), 7472–7479.

Skalova, T., Dohnalek, J., Ostergaard, L. H., Ostergaard, P. R., Kolenko, P., Duskova, J., et al. (2009). The structure of the small laccase from *Streptomyces coelicolor* reveals a link between laccases and nitrite reductases. *Journal of Molecular Biology, 385*(4), 1165–1178.

Spiker, J. K., Crawford, D. L., & Thiel, E. C. (1992). Oxidation of phenolic and non-phenolic substrates by the lignin peroxidase of *Streptomyces viridosporus* T7A. *Applied Microbiology and Biotechnology, 37*(4), 518–523.

Sugano, Y. (2009). DyP-type peroxidases comprise a novel heme peroxidase family. *Cellular and Molecular Life Sciences: CMLS, 66*(8), 1387–1403.

Sugiura, T., Yamagishi, K., Kimura, T., Nishida, T., Kawagishi, H., & Hirai, H. (2009). Cloning and homologous expression of novel lignin peroxidase genes in the white-rot fungus *Phanerochaete sordida* YK-624. *Bioscience Biotechnology, and Biochemistry, 73*(8), 1793–1798.

Suzuki, T., Endo, K., Ito, M., Tsujibo, H., Miyamoto, K., & Inamori, Y. (2003). A thermostable laccase from *Streptomyces lavendulae* REN-7: purification, characterization, nucleotide sequence, and expression. *Bioscience Biotechnology and Biochemistry, 67*(10), 2167–2175.

Thomas, L., & Crawford, D. L. (1998). Cloning of clustered *Streptomyces viridosporus* T7A lignocellulose catabolism genes encoding peroxidase and endoglucanase and their extracellular expression in *Pichia pastoris*. *Canadian Journal of Microbiology, 44*(4), 364–372.

Thurston, C. F. (1994). The structure and function of fungal laccases. *Microbiology, 140*(1), 19–26.

Tien, M., Kirk, T. K., Bull, C., & Fee, J. A. (1986). Steady-state and transient-state kinetic studies on the oxidation of 3,4-dimethoxybenzyl alcohol catalyzed by the ligninase of *Phanerocheate chrysosporium* Burds. *The Journal of Biological Chemistry, 261*(4), 1687–1693.

Updegraff, D. M. (1969). Semimicro determination of cellulose in biological materials. *Analytical Biochemistry, 32*(3), 420–424.

Valli, K., Wariishi, H., & Gold, M. H. (1990). Oxidation of monomethoxylated aromatic compounds by lignin peroxidase: role of veratryl alcohol in lignin biodegradation. *Biochemistry, 29*(37), 8535–8539.

Van Bloois, E., Torres Pazmiño, D. E., Winter, R. T., & Fraaije, M. W. (2010). A robust and extracellular heme-containing peroxidase from *Thermobifida fusca* as prototype of a bacterial peroxidase superfamily. *Applied Microbiology and Biotechnology, 86*(5), 1419–1430.

Vares, T., Niemenmaa, O., & Hatakka, A. (1994). Secretion of ligninolytic enzymes and mineralization of 14C-ring-labelled synthetic lignin by three *Phlebia tremellosa* strains. *Applied and Environmental Microbiology, 60*(2), 569–575.

Wang, Z., Bleakley, B. H., Crawford, D. L., Hertel, G., & Rafii, F. (1990). Cloning and expression of a lignin peroxidase gene from *Streptomyces viridosporus* in *Streptomyces lividans*. *Journal of Biotechnology, 13*(2–3), 131–144.

Wan, C., & Li, Y. (2012). Fungal pretreatment of lignocellulosic biomass. *Biotechnology Advances*.

Wariishi, H., Valli, K., & Gold, M. H. (1991). In vitro depolymerization of lignin by manganese peroxidase of *Phanerochaete chrysosporium*. *Biochemical and Biophysical Research Communications*, *176*(1), 269–275.

Watanabe, Y., Sugi, R., & Tanaka, Y. (1982). Enzymatic decolorization of melanoidin by Coriolus sp. No. 20. *Agricultural and Biological Chemistry*, *46*(6), 1623–1630.

Welinder, K. G. (1992). Superfamily of plant, fungal and bacterial peroxidases. *Current Opinion in Structural Biology*, *2*(3), 388–393.

Weng, J. K., Li, X., Bonawitz, N. D., & Chapple, C. (2008). Emerging strategies of lignin engineering and degradation for cellulosic biofuel production. *Current Opinion in Biotechnology*, *19*(2), 166–172.

Wesenberg, D., Kyriakides, I., & Agathos, S. N. (2003). White-rot fungi and their enzymes for the treatment of industrial dye effluents. *Biotechnology Advances*, *22*(1–2), 161–187.

Wong, D. W. (2009). Structure and action mechanism of ligninolytic enzymes. *Applied Biochemistry and Biotechnology*, *157*(2), 174–209.

Zheng, Z., Li, H., Li, L., & Shao, W. (2012). Biobleaching of wheat straw pulp with recombinant laccase from the hyperthermophilic *Thermus thermophilus*. *Biotechnology Letters*, *34*(3), 541–547.

Zimmerman, W., Umezawa, T., Broda, P., & Higuchi, T. (1988). Degradation of a non-phenolic arylglycerol β-aryl ether by *Streptomyces cyaneus*. *FEBS Letters*, *239*(1), 5–7.

CHAPTER TWO

Bacterial Volatiles and Diagnosis of Respiratory Infections

James E. Graham
Department of Microbiology and Immunology, and Department of Biology, University of Louisville,
Louisville, KY, USA
E-mail: j.graham@louisville.edu

Contents

Abstract

Breath testing has enormous potential in the medical diagnostic field. The underlying complexity and perceived availability of adequate specimens, combined with a lack of knowledge of the metabolic pathways that give rise to compounds that are sources of analytes detectable in breath, has greatly slowed development. These real obstacles have recently been largely overcome in the use of breath testing to identify patients with cystic fibrosis associated *Pseudomonas aeruginosa* infection and tuberculosis. This review summarizes progress made in the characterization of microbial volatiles produced by major lower respiratory tract bacterial pathogens, and their potential use as diagnostic markers in patient breath testing.

1. INTRODUCTION

Respiratory infectious diseases are among only a few common causes of patient visits to a doctor. The most prevalent, the common cold, is likely the most frequent human symptomatic infection (Eccles, 2005). Laboratory culture, a central aspect of the science of microbiology, remains the definitive tool in diagnosis of common bacterial respiratory infections in otherwise healthy patients. Nevertheless, culture and biochemical characterization are

Advances in Applied Microbiology, Volume 82
ISSN 0065-2164, http://dx.doi.org/10.1016/B978-0-12-407679-2.00002-8

no longer frequently used in routine medical practice simply because of the modest expense and minimal time associated with growth of bacteria that can be cultivated overnight in the laboratory media (Bartlett, 2011). Due to the typical low morbidity associated with the most common upper respiratory infections in otherwise healthy people, more often antibiotics are instead prescribed based on a patient history, doctor patient relationship, or self-report of sputum characteristics and length of symptoms (Tonkin-Crine, Yardley, & Little, 2011). Broad spectrum drugs are preferred, as most of the time the suspect bacterial agent is never confirmed. This contributes to acquisition of resistance in normal colonizers that can then be transferred to less benign residents and invaders by a variety of known mechanisms. Levels of use of antibiotics are directly related to levels of antibiotic resistance (Costelloe, Metcalfe, Lovering, Mant, & Hay, 2010), and an obvious goal is to reduce the inappropriate professional application of antibiotics in the treatment of respiratory infections.

A way to reliably identify a subset of agents causing respiratory infection in less than an hour at the site of patient contact would make a substantial contribution in this area (Ustianowski, 2012). Perhaps the most intuitively appealing approach to rapid point of care diagnosis of specific bacterial infections is analysis of patient breath, or "breath testing". It is now possible to identify infecting microorganisms based on volatiles they are known to produce. (In this summary of the current state of the field, breath testing will refer to detection of volatile metabolites in the breath, as opposed to "breath condensate" analysis, which should be considered as a distinct approach) (Hunt, 2011; Montuschi, 2007). While the proximity and accessibility of live microbes with distinct and diverse metabolic activities in or near airways provides a fundamentally important opportunity for efficient diagnosis, a series of challenges have greatly slowed use of the strategy. These include typically higher levels of both relevant environmental volatiles and potentially useful host-derived volatile metabolites in the same breath samples. Other obstacles include the need to concentrate and then efficiently recover volatiles present at very low levels for even the most sensitive of analyses, and that identical and very similar volatile compounds from different sources can contribute identical products to the analytical assays now used to identify them.

2. IDENTIFICATION OF VOLATILES AND VOCs

Volatile compounds are those whose vapor pressures at ambient temperatures are sufficient to escape from a liquid or solid surface and enter the air. For organic volatiles, these most often contain fewer than 20 carbons, and are under 500 Da (Skogerson, Wohlgemuth, Barupal, & Fiehn, 2011). Many are major secondary metabolites. All living things produce volatiles, but only a few have known roles in mediating ecological interactions. Volatile organic compounds or VOCs produced by plants are those best characterized, and include thousands of terpenoids, aromatic ring compounds, and amino and fatty acid derivatives (Baldwin, 2010). Several have known roles in pollination, defense, and fruit maturation (Kesselmeir & Staudt, 1999).

Gas chromatography (GC) together with electron impact (EI) mass spectrometry (MS) is the most established, reliable, and widely used approach to identify and characterize gasses and volatile compounds. Wider availability of this approach and the resolving power for different isomers in GC, and the existence of large reference databases and efforts at standardization (e.g. Kovats Index of n-alkane standards) (Phillips et al., 2012; Zhang et al., 2011) will increase the possibility for similar results to be obtained in different laboratories. Various methods of concentrating volatiles from reasonable volumes of air or breath (e.g. 1 L), including adsorbents (Woolfenden, 2010), solid-phase microextraction or SPME (Laaks, Jochmann, & Schmidt, 2011), and reactive coatings (Fig. 2.1) (Li, Biswas, Nantz, Higashi, & Fu, 2011; Strand, Bhushan, Schivo, Kenyon, & Davis, 2010), allow GCMS to be a sufficiently sensitive breath volatile detection method. Conventional high-temperature (>200° C) desorption on injection and "hard" ionization can complicate

Figure 2.1 Chemically reactive breath sampling plate used to characterize and monitor carbonyls from patient breath (Fu, Li, Biswas, Nantz, & Higashi, 2011; Li et al., 2011). (For color version of this figure, the reader is referred to the online version of this book.)

identification and reporting of novel volatiles and the definition of new puta-tive diagnostic markers (Spanel & Smith, 2011). Original compounds may be altered prior to ionization, and ions generated are not readily interpreted for microbial volatiles not in databases and lacking reference standards. Com-plex samples are thereby made more complex, and analyses lengthy. This approach will nonetheless likely continue to remain essential in defining volatiles, and indicating the target diagnostic markers for dedicated sensors or other more direct MS analyses more likely to be used in the point of care testing. Current successful methods (Chambers, Bhandari, Scott-Thomas, & Syhre, 2010) and the major analytical and statistical approaches used in breath testing have been recently reviewed (Boots et al., 2012; Buszewski, Kesy, Ligor, & Amann, 2007).

3. BACTERIAL VOLATILES

Our knowledge of VOCs produced by bacteria is quite limited com-pared to those released by plants. Although volatiles from different types of bac-teria have been characterized, bacterial VOCs for the most part do not appear in the major databases (i.e. Wiley, NIST, Massfinder) used for initial presumptive identification of compounds by comparison of mass spectra (Kai et al., 2009; Schulz & Dickschat, 2007). Fully defining more microbial volatiles will require novel methods to obtain sufficient quantities necessary for structural studies by isotope labeling and NMR. In our laboratories, we are currently using a covalent reactive capture approach with Fourier transform ion cyclotron mass spectrometry (FTICR-MS), together with GCMS, to define small vola-tile compounds (<500 Da) whose specific molecular formulae can be reliably defined in this way (Feng & Siegel, 2007).

Our knowledge of what bacterial volatiles and VOCs actually are and do is limited (Kai et al., 2009), and the corresponding metabolic pathways they derive from are mostly unknown. Volatiles are indeed abundantly produced, and characteristic of different types of bacteria. For exam-ple, *Escherichia coli* growing on the conventional yeast extract-tryptone agar produces large amounts of the volatile indole. Actinomycete bacteria in the genus Streptomyces are abundant in soil, and produce geosmin and related volatiles. Cultures of *Pseudomonas aeruginosa*, discussed below, often have a distinctive "grape-like" odor due to the production of 2-amino-acetophenone (Scott-Thomas et al., 2010), which has demonstrated good potential as a volatile breath marker of respiratory infection.

Identification of bacterial volatiles typically involves growth in laboratory cultures where production of different volatiles is dependent on type of media, time of incubation, amount of aeration, and other environmental variables (Carey et al., 2011; Chippendale, Spanel, & Smith, 2011; Nawrath, Mgode, Weetjens, Kaufmann, & Schulz, 2012; Preti et al., 2009). These parameters can also influence our ability to detect different volatile chemicals produced by the bacteria, which although largely stable can be reactive (e.g. aldehydes and ketones) and of different solubilities and likelihood of being adsorbed onto surfaces. For the most reproducible mid-logarithmic phase laboratory broth cultures, only volatiles capable of escaping broth culture media with reasonable Henry partition constants ($K_{air/water}$) can reach detectable headspace levels (in the nanomolar or 0.1 ppm range). The different media typically used in the culture of bacteria are most often derived from biological sources, and produce their own characteristic volatiles, which may be increased or altered by autoclaving, or made more volatile by bacterial metabolic activity or physical alteration of agar surfaces (Schulz & Dickschat, 2007). Experiments designed to identify bacterial volatiles then must identify those released from media and subtract levels for what may appear to (and may) be identical volatiles by available analytical methods. Headspace air above the cultures can be collected by a variety of means, and for laboratory cultures a custom "closed loop stripping apparatus" is often constructed to allow volumetric determination from a hermetically closed culture system. Alternately, headspace air may be sampled directly by MS approaches that do not rely on prior separation of compounds by chromatography (Boots et al., 2012). Bacterial culture volatiles may also be adsorbed onto different surfaces or "trapped" in inherently less quantitative but still highly useful approaches, as described in the numerous primary studies cited below.

It is unlikely that breath volatile makers originating from bacteria important in respiratory and other infections will become mainstream and reliable indicators of disease until their metabolic origins and chemical structures are fully defined using isotopes and laboratory cultures. We also may not find those of actual utility unless we also look simultaneously at microbes growing in their normal ecological niches. For the bacteria of high relevance to respiratory infection, these are of course the host cells and tissues in which they normally reside.

4. VOLATILES FROM MAJOR BACTERIAL RESPIRATORY PATHOGENS

Bacteria cause roughly a third to a half of adult community-acquired respiratory infections (Fine et al., 1996; Johansson, Kalin, Tiveljung-Lindell, Giske, & Hedlund, 2009; Lim et al., 2001). These are often secondary to viral infections, warranting consideration of the use of antibiotics in only about half of cases. Current Medicaid guidelines in the USA call for administration of antibiotics within 6 h for community-acquired pneumonias (File, Solomkin, & Cosgrove, 2011). A few types of bacteria are of very high specific interest with regard to the common adult community-acquired lower respiratory infections. These include *Streptococcus pneumonia* and *Haemophilus influenzae,* both of which are often present in the upper respiratory tract of healthy individuals *Moraxella* (formerly *Branhamella*) *catarrhalis* is only normally prevalent in children, frequently causing ear infections, but is also being increasingly recognized as an as a contributor in adult pneumonia (Ramirez & Anzueto, 2011). *Mycobacterium tuberculosis* is estimated to currently inhabit about a third of the world's population, most often resulting in a life long asymptomatic infection. This becomes of enormous importance for respiratory disease diagnosis when these bacteria instead grow to high levels in the extracellular spaces or cavities in the lung that are created by a delayed-type hypersensitivity (DTH) immune response. While not commonly seen in community-acquired infections, *P. aeruginosa* and *Burkholderia cepacia* are of high interest in terms of colonization of the lung in cystic fibrosis, the most common genetically inherited disease in most countries. For all of these pathogens, there is at least one report of volatile compounds produced during growth in different environments, and Table 2.1 summarizes these data. *M. tuberculosis* and *P. aeruginosa* have been more extensively characterized, and breath testing has already been used in individual trials to identify patients infected by these microbes and is discussed below.

S. pneumonia is by far the most prevalent and important cause of significant lower respiratory tract infections among otherwise healthy individuals. A recent meta-analysis of patient data indicates that it is most likely that a recent unusual drop in the reported frequency as the causative agent of community-acquired pneumonia (CAP) is due to a dramatic failure in continuing to identify specific bacterial or other agents causing respiratory disease (Bartlett, 2011). While there is a great deal of information on volatile components produced by related Streptococci used in food and beverage fermentation, very few studies have attempted to characterize those produced by the pneumococcus. It appears none have yet analyzed the potential contributions of this diverse microbe (Hiller et al., 2007) to human breath.

Growth of these bacteria in conventional clinical roller bottle media was shown to produce different levels of acetaldehyde, acetone, and ethanol in culture headspace (Allardyce, Langford, Hill, & Murdoch, 2006), but no unique compounds were identified. In contrast, Ishimaru et al. (Ishimaru, Yamada, Nakagawa, & Sugano, 2008) identified numerous analytes including many potentially unique to cultured *S. pneumoniae* by the soft ionization MS approach APCI-MS (atmospheric pressure chemical ionization). Their sensitive and high resolution approach did not allow identification of the corresponding volatiles. In contrast, a recent carefully calibrated approach to sampling volatiles produced by *S. pneumonia* growing on sheep's blood agar using multi-capillary column–ion mobility spectrometry (MCC-IMS) found only one unidentified *S. pneumonia* volatile compound, but it was also detected above plates growing 10 of 14 other potential bacterial respiratory pathogens (Junger et al., 2012). A recent important study of volatiles produced by *S. pneumoniae* and other upper respiratory pathogens frequently cultured from infected nasal mucus by Preti et al., (2009) showed that *S. pneumoniae* volatiles produced in culture and in vivo included both common bacterial volatiles and unique compounds. In vitro, *S. pneumoniae* produced benzaldehyde, benzylalcohol, 2-phenylethyl alcohol, acetic acid, and methyl mercaptan, all of which were produced by *H. influenzae* and *M. catarrhalis* with the exception of methyl mercaptan, which has been identified in other studies as produced by *P. aeruginosa* (Table 2.1). *H. influenzae* and *M. catarrhalis* each produced a subset of volatiles produced by *S. pneumoniae* on this agar, and no distinct volatiles for these isolates were detected by the authors' approach. Difficulty in distinguishing volatiles produced by *S. pneumoniae* and *P. aeruginosa* grown on plates and resuspended in buffer was previously noted using a polymer composite sensor array (or electronic nose) that was able to distinguish among other major bacteria in lower respiratory infection (*Staphylococcus aureus, S. pneumoniae, H. influenza,* and *P. aeruginosa*), and these pathogens and control samples lacking bacteria (Lai, Deffenderfer, Hanson, Phillips, & Thaler, 2002). Taken together, these few studies suggest that a breath test for known microbial volatiles intended to determine "bacteria positive or negative" is more likely than one able

Table 2.1 Volatile organic compounds produced by microorganisms in standard culture media

Respiratory bacteria	Identified volatiles	Studies
Streptococcus pneumoniae	Ethanol, benzylalcohol, 2-phenylethanol, aminoacetophenone acetaldhyde, benzaldehyde, acetic acid, indole, methyl mercaptan, dimethyl sulfide, trimethylamine, hexanal,	*a, e*
Haemophilus influenzae	Benzaldehyde, benzylalcohol acetic acid, indole	*a*
Moraxellacatarrhalis	Benzaldehyde, benzylalcohol, and 2-phenylethanol	*a*
Pseudomonas aeruginosa	Ethanol, 1-butanol, 1-pentanol, isopentanol, 2-amino-acetophenone, 2-butanone, 2-nonanone, 3-undecanone, isoprene, 1-undecene, hydrogen sulfide, methyl mercaptan, dimethyldisulfide, dimethyltrisulfide, dimethylpyrazine hydrogen cyanide, methyl thiocyanate, formaldehyde, ammonia, trimethylamine	*a, b, c, d, e, i*
Mycobacterium tuberculosis	1-hexanol, 2-phenylethanol, 2-hydroxy 3-pentanone, g-methyl butyrolactone, methyl nicotinate★, methyl para-anisate★, ortho-phenylanisole★, methyl phenylacetate★, camphor + also those in refs Phillips et al. (2010), Phillips et al. (2007), Nawrath et al. (2012), Kolk et al. (2012)	*g, h*
Burkholderia cepacia	Ethanol, 1-pentanol, 2-aminoacetophe-none, ethylbutanoate, hydrogen sulfide, methyl mercpatan, isoprene, acetoin, ethyl acetate, acetic acid, butanoic acid, phenyl acetic acid, pyrrol, formaldehyde, ammonia, trimethlamine	*f*

Volatiles detected above the background levels from the media by various approaches in different laboratories.
★These four potentially *M. tuberculosis*-unique aromatic ring volatiles were original identified by Syhre & Chambers, (2008), and recently independently confirmed by Mgode, Weetjens, Cox, et al. (2011). *a* (Preti et al., 2009), *b* (Labows, McGinley, Webster, & Leyden, 1980), *c* (Zechman, Aldinger, & Labows, 1986), *d* (Carroll et al., 2005), *e* (Allardyce et al., 2006), *f* (Thorn et al., 2010), *g* (Syhre & Chambers, 2008), *h* (Mgode, Weetjens, Nawrath, et al., 2011), *i* (Shestivska et al., 2011).

to be able to distinguish among the three most prevalent "typical" bacterial lower respiratory pathogens. Identifying these three will more likely require considering more than a single or small group of defined microbial marker volatiles, as is currently being done with the pattern recognition and sensor arrays, and statistical analysis of data on many analytes from mass spectra (Fend et al., 2006; Phillips et al., 2012; Robroeks et al., 2010; Thorn, Reynolds, & Greenman, 2010).

A great deal of effort to provide a description of the limited number of known bacterial metabolic pathways that generate volatile compounds has been put forth by Schulz and Dickschat (2007). Those described are now included in an MS database known as Superscent (Dunkel et al., 2009). A recent review by Thorn and Greenman (2012) gives an excellent broad overview of the central pathways leading to major classes of known bacterial volatiles currently of interest in developing respiratory infection diagnostics. It will only be through understanding exactly what respiratory bacterial diagnostic maker volatiles are, and how and when they are generated during infection, that we will be able to use them as targets to develop highly reliable breath tests for the most common bacterial respiratory infections.

5. DEVELOPMENT OF BREATH TESTING

In contrast, volatiles found in human breath are rather well characterized. Hippocrates is said to have used breath odor as a means of diagnosing patient disease (Geist, 1957). The possibility of obtaining and analyzing human exhaled air for its volatile chemical components by GC was first demonstrated 50 years ago by Linus Pauling, who trapped breath in 5 ft-long coiled stainless steel tubes or "traps" (Pauling, Robinson, Teranishi, & Cary, 1971; Teranishi, Mon, Robinson, Cary, & Pauling, 1972). These were cooled to freeze and remove excess water and all "condensate" (see below) prior to analysis. A heat gun was then used to immediately "flash" air in the metal tube into a GC column for analysis by an ion flame detector. The initial experiments showed that normal breath gases and volatiles produced about 250 analytes. We now know these are primarily from nitrogen, oxygen, carbon dioxide, water vapor, and much lower amounts of VOCs including acetone, isoprene, ethane, and pentane (Kwak & Preti, 2011; Miekisch, Schubert, & Noeldge-Schomburg, 2004; Whittle, Fakharzadeh, Eades, & Preti, 2007). Even abundant VOCs are present at only parts per billion (ppb, approximately nMol) levels, and others likely of utility in diagnosis at parts per trillion (ppt, approximately pMol) concentrations. Detection at such low levels has been achieved by different MS approaches, where sensitivity (Kwak & Preti, 2011) and the use of widely different and quickly evolving analytic methods (Thorn & Greenman, 2012) have resulted in very little overlap in reports of potentially useful breath test biomarkers by different research groups. As demonstrated by Boshier, Marczin, & Hanna (2010) with quantitative selective ion flow tube or SIFT-MS directly on exhaled breath, relative levels of normal breath VOCs can be reliably assessed in individual laboratories. Reasonably sensitive diagnosis of *P. aeruginosa* (Savelev et al., 2011; Scott-Thomas, Pearson, & Chambers, 2011; Scott-Thomas et al., 2010), *M. tuberculosis* (Kolk et al., 2012; Phillips et al., 2012, 2010, 2007; Syhre, Manning, Phuanukoonnon, Harino, & Chambers, 2009), and *Aspergillus fumigatus* (Chambers, Syhre, Murdoch, McCartin, & Epton, 2009) lung infections by the patient breath testing has also recently been demonstrated, as described below.

A second major challenge that has slowed the development of diagnostic breath testing is that most human breath VOCs and many proposed disease biomarkers are present in the environment. Breath is composed of a mixture of endogenously produced and environmental compounds. Early studies by Michael Phillips, a leader in this area, proposed the calculation of an "alveolar gradient" (Phillips, Greenberg, & Awad, 1994), or identification and quantification of local ambient air volatile levels to "subtract" from levels found in the exhaled breath. While clearly a required consideration, others have since pointed out (Bajtarevic et al., 2009; Kwak & Preti, 2011) that inspired volatile compounds can be concentrated in the body, and therefore higher levels in the breath relative to the environment do not necessarily indicate an endogenous origin of an identical compound. With volatiles of potential diagnostic interest present in both the environment and produced by healthy patients, breath volatile quantification becomes one of paramount importance. Unfortunately, this is not a strength of the most widely used sampling and analysis methods. Nonetheless, numerous studies are now available that clearly establish that individual specific volatiles or combinations of detected MS analytes in breath analyses can be used with good accuracy to identify diseases, including diabetes, asthma, breast and lung cancer, cystic fibrosis, COPD, and liver disease (reviewed by (Buszewski et al., 2007; Whittle et al., 2007)). Standardizing approaches across different laboratories (as described below) and then bringing breath testing to respiratory infection patients for point of care testing will likely require further reducing the cost of tests by a second level of development (Smith, 2011). There is considerable potential for contributions in the form of dedicated sensor technology (e.g. "electronic noses") capable of monitoring individual previously identified or unknown volatiles in complex mixtures (Lai et al., 2002). Sensor approaches, like soft ionization analyses with SIFT-MS and proton transfer reaction spectrometry (PTR-MS), are also capable of defining temporal patterns of evolution of identified markers (Bunge et al., 2008; O'Hara & Mayhew, 2009) of diagnostic value directly in breath.

Defining changes in endogenous volatile compounds in breath that result from infection rather than microbial volatiles is also likely to contribute to our ability to diagnose bacterial respiratory infections by breath testing. Although endogenous markers are not ideal in terms of contributing specificity across patient populations, their inherently higher levels may make their use in breath testing necessary. Alterations of better characterized human volatile metabolites in the breath do occur with respiratory disease, including infections. These changes likely result from the different and specific types of inflammation and temporal recruitment of different populations of immune cells producing different factors and cell signaling molecules. All pathogens entering the alveoli first encounter resident alveolar macrophages capable of secretion of proinflammatory cytokines including interleukin (IL)-8, IL-1β, and tumor necrosis factor (TNF)-α. Inflammation associated with pneumococcal infection, for example, is characterized by the central role of the innate immune response and influx of neutrohpils (Delclaux & Azoulay, 2003). The potential for a characteristic pattern of breath volatiles to reflect this kind of immediate and innate inflammatory response has been demonstrate recently in a rare animal model of breath testing with Sprague–Dawley rats (Guaman et al., 2012). In contrast, *M. tuberculosis* is able to limit proinflammatory signaling when encountering alveolar macrophages not previously activated by other stimuli. If able to survive phagocytosis, local inflammation is then initially limited for a time until a DTH develops, characterized by infiltration of cells with central roles in acquired immunity (T lymphocytes, including regulator T-cells, and B-cells). This results in a very different granulomatous inflammation and remodeling of lung tissues and may eventually result in active disease and large scale tissue destruction (including cavity formation, with high levels of extracellular bacteria directly in the airway). Differences in metabolism among relevant cells and different populations of cells are then expected to lead to changes in endogenous volatile production in the breath (Mashir et al., 2011), as now well described for other noninfectious diseases (Whittle et al., 2007).

The fact that lung tissue inflammation is characteristic of tuberculosis is reflected in the evolution of *M. tuberculosis* to use signals encountered in environments rich in reactive oxygen and nitrogen to regulate expression of factors important for adaptation (Trivedi, Singh, Bhat, Gupta, & Kumar, 2012). Volatile carbonyl compounds and saturated hydrocarbons that are abundant in breath (e.g. acetone, ethane, and pentane) are thought to arise from membrane lipid and fatty acid peroxidation (Miekisch et al., 2004). Interestingly, this resembles the release of green leaf volatiles from mechanically injured plants (Baldwin, 2010). An important early study by Phillips (Phillips et al., 2007) showed a previously established index of normal breath alkanes thought to be generated by lipid peroxidation did not increase in *M. tuberculosis* culture positive respiratory patients compared to patients hospitalized with other respiratory diseases. A group of 12 other breath markers were significantly associated with culture positive status, and potentially reflect differences in lung inflammation characteristic of infection by *M. tuberculosis*, in addition to a smaller of number volatiles that may have originated directly from the bacteria. Robroeks et al. (2010) also used a combination of 14 volatile markers likely to have included those originating from the patient's response to infection to achieve 100% correct classification of 23 *P. aeruginosa* culture positive among 48 pediatric CF respiratory patients.

6. DIAGNOSIS OF RESPIRATORY INFECTION BY ANALYSIS OF VOLATILES

Breath testing is probably the least developed of any recognized area of medical diagnostic testing. It is also perhaps both the least invasive and most readily accepted by the patient. In terms of identifying bacterial respiratory diseases, breath testing is clearly an area of extremely high potential. The only frequently used breath test for a microbial infection involves *H. pylori,* which specifically colonies the gastric mucosa rather than the respiratory tract. This application is also not entirely noninvasive, as detection of *H. pylori* urease activity, a metabolic capability shared by many microbes (Jassal et al., 2010) but absent from human hosts, currently involves administration of radiolabeled urea and detection of labeled carbon dioxide. *H. pylori* colonization is also reported to elevate levels of both breath carbon dioxide (Pathak, Bhasin, & Khanduja, 2004) and nitrate and cyanide (Lechner et al., 2005). In contrast to most areas of the life science research, the use of animal models of infection (Guaman et al., 2012; Purkhart et al., 2011) has made only very limited contributions to progress in this area, likely because of the key role of specific environment in microbial volatile production previously described. Successful experimental human breath testing for *P. aeruginosa* and *M. tuberculosis* and the fungal respiratory pathogen *Aspergillus* has recently been demonstrated. *A. fumigatus* breath testing has been reviewed recently by Chambers and and is not summarized here (Chambers et al., 2010).

M. tuberculosis and *P. aeruginosa* are the two lower respiratory pathogens that now appear to be the most amenable to detection by breath testing. Interestingly, they are both known for their metabolic capacities and flexibility, having large (~4,000 open reading frame or ORF) genomes and very large (6,000 ORF) genomes, respectively. While *Pseudomonas* normally lives in a variety of environmental reservoirs and is an opportunistic pathogen, *M. tuberculosis* normally inhabits only human cells and tissues. A large genome for an obligate intracellular pathogen provides the reserve capacity needed to live in a variety of different human cells and tissues at different stages of pulmonary infection (Bishai, 1998). Neither of these bacteria is ever considered part of normal flora colonization or "microbiota" of healthy individuals. This increases our ability to detect their presence from the volatiles they produce and evoke from the host during respiratory infection.

Infection by *P. aeruginosa* is only infrequently associated with CAP (Fine et al., 1996) but is instead one of the most important causes of hospital acquired lower respiratory infection. Children with cystic fibrosis and patients with immunocompromise or chronic lung conditions including chronic obstructive pulmonary disease or COPD (Fujitani, Sun, Yu, & Weingarten, 2011) are most frequently infected. *P. aeruginosa* infection rates are much higher in the hospital setting where they impact patients with chronic lung disease and HIV. In the U.S. hospital intensive care units, *P. aeruginosa* was the most frequently (31.6%) isolated bacterium from respiratory culture (Neuhauser et al., 2003). These infections are often associated with use of ventilators and bronchoscopes.

Enderby, Smith, Carroll, and Lenney (2009) reported an early effort to use SIFT-MS detection of previously reported *P. aeruginosa* volatile hydrogen cyanide to identify infected CF patients. HCN levels were found to be significantly elevated in 16 culture positive patients among 37 pediatric respiratory patients, but the authors found results were not consistent over 6 months in retesting. Hydrogen cyanide and methyl thiocyanate were found to be reliably produced by strains of *P. aeruginosa* in culture but did not prove useful as breath markers as more recently reported by Shestivska et al. (2011).

Robroeks et al. (2010) were the first to describe a comprehensive evaluation of potential breath marker volatiles and their successful use in diagnosis *P. aeruginosa* infections among CF patients. They used what is now a classic approach of collecting breath into 5 L polycarbonate (e.g. Teldar) bags, followed by drawing the contents through carbon adsorbant containing tube. GC Time of flight (TOF) MS was then used to identify ions and a group of 14 marker analytes by statistical discriminant analyses. This combination of 14 markers was able to discriminate 100% of the 23 *P. aeruginosa* culture positive patients among 39 CF patients.

The aromatic ketone 2-acetoaminophenone, which imparts a grape-like odor to *P. aeruginosa* cultures, has been described (Scott-Thomas et al., 2010) as a sensitive and specific (93.8% and 69.2%, respectively) indicator of *P. aeruginosa* culture positive status. Although also previously identified in headspace analyses of *S. pneumonia* growing in BacTec roller bottles (Allardyce et al., 2006; Scotter, Allardyce, Langford, Hill, & Murdoch, 2006), the authors did not identify the volatile as produced by cultures of this or other bacteria (several strains of *H. influenzae*, *Legionella pneumophila*, *M. catarrhalis*, *Pseudomonas fluorescens*, and *Burkholderia multivorans*) in their assay. 2-acetoaminophenone demonstrated high utility as a biomarker in identifying 15 of 16 *P. aeruginosa* infected CF patients among 46 total participants, including other CF patients, many with *S. aureus* infections. The authors used breath collection into 1 L glass bulbs (after finding 2-acetominophene unstable in Teldar polycarbonate bags), into which were inserted adsorbant SPME needles, and the compound was detected by targeted CG-MS-MS.

The largest studies reporting successful patient breath testing for bacterial respiratory infections to date come from efforts to diagnosis tuberculosis. Diagnosis of *M. tuberculosis* by breath testing is of great specific interest as it is only in active disease when bacteria grow in extracellular spaces within the lung (Grosset, 2003) that tuberculosis patients are infectious. If these cases can be efficiently recognized in this way, new cases could be dramatically reduced in the remaining high prevalence regions. That *M. tuberculosis* produces recognizable volatiles when growing in the lung is clearly demonstrated by the ability of both honeybees (Suckling & Sagar, 2011), and trained African giant pouch rats (Mgode, Weetjens, Cox, et al., 2011; Weetjens et al., 2009) to identify positive sputum specimens. Recently specific volatiles including three diterpenes have been described (Prach, Kirby, Keasling, & Alber, 2010). Iso-tuberculosinol (originally called Edaxadiene) has been shown to block phagosome maturation in infected macrophages and is only produced in standard media when magnesium levels are reduced (Mann et al., 2009). *M. tuberculosis* also has an extraordinary metabolic capability to synthesize and degrade lipids, and constituent fatty acids are precursors to many different known volatiles including alkanes and carbonyls. *M. tuberculosis* volatile metabolite production changes with culture on different media, and has now been characterized by Nawrath et al. (2012).

In the first reported effort at testing patient breath to diagnose tuberculosis (Phillips et al., 2007), researchers obtained filtered patient-breath specimens from 42 hospitalized potential tuberculosis patients in New York with a portable multistage breath collection device capable of sampling a subset of expired air (Phillips, 1997; Phillips & Greenberg, 1992). The device continuously collects breath samples from which volatiles are adsorbed onto a conventional carbon matrix. Volatiles adsorbed from 1 L of breath (2 min) were thermally desorbed by heating in a stream of helium and the volume of collected gas then reduced by two sequential cyrofocusing traps, prior to analysis of the entire sample by conventional GCMS. Room air values were also obtained and subtracted for each spectra. Results from 59 local age matched healthy volunteers were compared to those from 23 culture positive tuberculosis patients. Statistical multivariate analyses were then used to identify a subset of 134 analyte ions identified for each sample. Principal component analysis indicated that a group of 10–12 identified analytes could be used as a test with a sensitivity of 82.6% for culture positive patients, and 100% specificity. (Current widely used sputum smear microscopy was estimated to detect only 63% of cases as of 2007) (WHO, 2010). Two different types of statistical analyses on these data indicated two different groups of putative breath marker volatiles, with both including multiple volatiles related to heptane and benzene (Phillips et al., 2007). Comparison of 1 ml of headspace from *M. tuberculosis* cultures grown in commercial VersaTREK bottles by the same breath collection and analysis approach indicated just 1-methyl-naphthalene and 1,4-dimethyl-cyclohexane as candidate bacterial derived volatiles that could be identified from both laboratory cultures and in culture positive patient breath.

A larger multisite (the U.S., U.K., and Philippines) application of the same approach by these investigators (Phillips et al., 2010) analyzed breath volatiles in 226 symptomatic high risk patients. Markers of good predictive value were identified including both proposed host oxidative stress products (alkanes and alkane derivatives) and the previously described volatile *M. tuberculosis* metabolites cyclohexane and 1,3,5-trimethyl-benzene. Researchers were able to identify active pulmonary tuberculosis with 84% sensitivity and 64.7% specificity when sputum culture, microscopy, and chest radiography were either all positive or all negative, and 70% specificity when using positive sputum culture as the only criterion for infection.

A smaller pilot study by Syhre et al. (2009) also assayed 10 smear positive patient specimens and 10 healthy controls by collecting breath into 1 L silanized glass bulbs, which were sampled with adsorbant coated SPME needles. Analytes were thermally desorbed directly into a CGMS-MS and scanned specifically to identify methyl nicotinate, one of the four previously identified culture volatiles (methyl para-anisate, ortho-phenylanisole, methyl phenylacetate), now confirmed by Mgode et al. (2011). Detection of methyl nicotinate were significantly higher in subjects with tuberculosis, but still detectable in the breath of healthy controls.

A good example of how quickly this area is now moving is that during the preparation of this review, two new tuberculosis patient breath-testing studies have been reported. Kolk et al. (2012) studied patients in the endemic setting in South Africa, where 28% of culture proven TB patients had Ziehl-Neelsen (ZN) negative sputum smear. Breath was collected by normal breathing into Teldar bags, and analytes collected by drawing the contents through an adsorbant carbon trap. No environmental sampling was used to remove contributions from room air, and no specific breathing maneuvers were performed. An initial set of breath samples from 50 known sputum culture positive and 50 culture negative patients was analyzed using GC TOF-MS. Using six putative volatiles (dodecane, 3-Heptafluorobutyroxypentadecane, 5-hexenoic acid, 2-ethyl-1-hexanol, tetradecanoic acid, octanal) and one unknown analyte identified as markers, testing on an independent group of 21 TB and 50 non TB patients from the same area gave a sensitivity of 62%, and a specificity of 84%.

Phillips et al. (2012) have refined their approach to include a portable gas chromatograph coupled with surface acoustic wave (SAW) detector, achieving a major goal in allowing the point care testing. Their previously described (above) approach was used to collect breath from 279 patients in the U.K., Philippines, and India (130 active pulmonary TB, 121 controls), and the results collected and analyzed remotely by internet connection. The authors used an important standardization by modified Kovat's Indices to reduce differences at different sites of deployment. Analytes corresponding to the previously described and novel volatiles camphene, camphene, L-beta-pinene, 1,3,5-trimethyl-benzene, 1-methyl naphthalene, tridecane, 2-butyl 1-octanol, 4-methyl dodecane, and 2,2,4,6,6-pentamethyl-heptane were combined into 4 co-resolving regions in their assay system and scored, indicating 71.2% sensitivity and 72% specificity in identifying active tuberculosis cases among the same specimens used to identify markers.

7. STANDARDIZATION

After an initial delay of perhaps 35 years since the original description of breath testing for volatiles, it now seems clear that we can identify at *least M. tuberculosis* and *P. aeruginosa* lower respiratory infections by patient breath testing. More common bacterial agents of community-acquired adult pneumonia are proving more difficult to diagnose, likely because they are both part of and perhaps more similar to other bacteria of the normal upper and lower respiratory tracts. Our knowledge of the production of volatiles by these bacteria in the laboratory is also more limited. What caused this delay, and what did we learn from it?

As previously described, breath testing for volatiles is not the same as "breath condensate" analysis (Montuschi, 2007). Breath condensates are far more complex human biological specimens that contain an even broader range of compounds, including nonvolatile leukotrienes, peptides, and cytokines. Condensate are also considerably easier to obtain specimens (Hunt, 2011), and so a greater number of different specific approach to collection, storage, and concentration (Rosias, Robroeks, Hendriks, Dompeling, & Jobsis, 2004) have added variability to the already diverse application of very different MS techniques. As with the analysis of breath volatile compounds as described in this review, accurate quantification of changes in endogenously produced components is a central aspect. It is more difficult with condensates because they include aerosolized nonvolatiles escaping from the airway epithelial surface, and breath is saturated with water vapor (Effros, 2010).

The same problems of quantification and question of origins are present in breath testing for volatiles. Within a particular patient breath specimen, no volatile compound has been identified that can be used as an internal standard to evaluate levels of other compounds. Standards can be added exogenously to the collected air or adsorbed desorbed sample, but then the amount and origin of breath represented by the specimen is not standardized. Variability in collection is expected, particularly as there are typically respiratory issues among those needing diagnosis. Many studies have attempted to address the issue of variability in collection of breath. Issues in standardizing breath sampling that occur before analytes are collected are the same for analyses of breath volatiles and condensates, and have been thoroughly described previously in the context of obtaining breath condensates (Horvath et al., 2005; Rosias, 2012). It does appear that many differences in the manner and environment in which breath is collected have the potential to impact the quantification of a subset of volatiles, and those will perhaps not be useful as diagnostic markers. However, there are many potential markers and methods to increase reliability in obtaining samples. These include avoiding a 10–20% dilution from the initially evacuated upper respiratory air in each breath by "end tidal" sampling, which can potentially improve sampling reliability (Thekedar, Oeh, Szymczak, Hoeschen, & Paretzke, 2011). A laboratory technique of pre breathing pure bottled air (Basanta et al., 2012) can also reduce quantification issues in some settings. There is also the possibility of using only multiple ratios of detected analytes within each specimen as a potentially recognizable matrix of values for statistical consideration.

In terms of identifying the origins of genuine volatiles in the breath, this is a key question in that only by determining the metabolic sources of volatile markers, microbial or host, are we likely to move the field forward and develop superior and less invasive diagnostic tests. As described above, many low molecular weight volatiles are fairly ubiquitous from biological sources, as are all of those so far identified derived from the host whose levels may change with different immune responses. Microbial volatiles present in the body may be metabolically altered by the host and not recognizable in the breath. In the lung, diffusion of volatiles from the blood to the alveolar air across the alveolar capillar membrane also depends on other volatile chemical properties, including polarity and solubility in fat, and will not be the same as those detected in headspace of blood and urine specimens (King et al., 2010, 2011). This can perhaps only be addressed with cell culture and animal models using labeled isotopes— work now being carried out in the author's laboratory.

8. THE FUTURE OF DIAGNOSIS BY BREATH TESTING

It now seems likely that we will see clinical diagnostic breath tests first for tuberculosis, and then perhaps nosocomial and community-acquired infections by *P. aeruginosa*. A breath test able to detect any bacterial agent relative to viral etiology appears somewhat further off, but would likely see even wider application. Future development of simplified sensor tests will likely benefit next from the experimental use of a combination of comparative direct analyses of just easily obtained "mixed expiratory" breath, for example, identification of bagged breath volatiles by SPME and GCMS (qualitative), and confirmation and quantification directly in breath by SIFT-MS (quantitative). Once biomarkers have been identified and confirmed, targeted and simplified MS (Phillips et al., 2012) or customized sensors arrays can be designed for the point care testing. These are most likely to be used in combination with other diagnostic tests, to 'triage', or rule in or out specific infections. It is likely that the ability to obtain and use of temporal patterns (Bunge et al., 2008; Carey et al., 2011) perhaps in combination with an administered microbial substrate (Jassal et al., 2010), as done with diagnosis of gastrointestinal disorders (Saad & Chey, 2007), will help to alter otherwise similar volatiles originating from normal colonizers and those responsible for the bacterial respiratory disease.

ACKNOWLEDGMENTS

I wish to thank my University of Louisville colleagues Dr Xiao-An Fu (Department of Chemical Engineering) who made the breath volatile reactive sampling plate technology we are using, and Drs Richard Higashi and Michael Nantz (Department of Chemistry) for challenging me with the complexities of mass spectrometry and volatile analytical and organic chemistry. Our work in this area is supported by the Global Health initiative of the Bill and Melinda Gates Foundation and by the National Institutes of Health.

REFERENCES

Allardyce, R. A., Langford, V. S., Hill, A. L., & Murdoch, D. R. (2006). Detection of volatile metabolites produced by bacterial growth in blood culture media by selected ion flow tube mass spectrometry (SIFT-MS). *Journal of Microbiological Methods, 65*, 361–365.

Bajtarevic, A., Ager, C., Pienz, M., Klieber, M., Schwarz, K., Ligor, M., et al. (2009). Noninvasive detection of lung cancer by analysis of exhaled breath. *BMC Cancer, 9*, 348.

Baldwin, I. T. (2010). Plant volatiles. *Current Biology, 20*, R392–R397.

Bartlett, J. G. (2011). Diagnostic tests for agents of community-acquired pneumonia. *Clinical Infectious Diseases, 52*(Suppl. 4), S296–S304.

Basanta, M., Ibrahim, B., Douce, D., Morris, M., Woodcock, A., & Fowler, S. J. (2012). Methodology validation, intra-subject reproducibility and stability of exhaled volatile organic compounds. *Journal of Breath Research, 6*, 026002.

Bishai, W. (1998). The *Mycobacterium tuberculosis* genomic sequence: anatomy of a master adaptor. *Trends in Microbiology, 6*, 464–465.

Boots, A. W., van Berkel, J. J., Dallinga, J. W., Smolinska, A., Wouters, E. F., & van Schooten, F. J. (2012). The versatile use of exhaled volatile organic compounds in human health and disease. *Journal of Breath Research, 6*, 027108.

Boshier, P. R., Marczin, N., & Hanna, G. B. (2010). Repeatability of the measurement of exhaled volatile metabolites using selected ion flow tube mass spectrometry. *Journal of the American Society for Mass Spectrometry, 21*, 1070–1074.

Bunge, M., Araghipour, N., Mikoviny, T., Dunkl, J., Schnitzhofer, R., Hansel, A., et al. (2008). On-line monitoring of microbial volatile metabolites by proton transfer reaction-mass spectrometry. *Applied and Environmental Microbiology, 74*, 2179–2186.

Buszewski, B., Kesy, M., Ligor, T., & Amann, A. (2007). Human exhaled air analytics: biomarkers of diseases. *Biomedical Chromatography, 21*, 553–566.

Carey, J. R., Suslick, K. S., Hulkower, K. I., Imlay, J. A., Imlay, K. R., Ingison, C. K., et al. (2011). Rapid identification of bacteria with a disposable colorimetric sensing array. *Journal of American Chemical Society*, *133*, 7571–7576.

Carroll, W., Lenney, W., Wang, T., Spanel, P., Alcock, A., & Smith, D. (2005). Detection of volatile compounds emitted by *Pseudomonas aeruginosa* using selected ion flow tube mass spectrometry. *Pediatric Pulmonology*, *39*, 452–456.

Chambers, S.T., Bhandari, S., Scott-Thomas, A., & Syhre, M. (2010). Novel diagnostics: progress toward a breath test for invasive *Aspergillus fumigatus*. *Medical Mycology*, *49*(Suppl. 1), S54–S61.

Chambers, S. T., Syhre, M., Murdoch, D. R., McCartin, F., & Epton, M. J. (2009). Detection of 2-pentylfuran in the breath of patients with *Aspergillus fumigatus*. *Medical Mycology*, *47*, 468–476.

Chippendale, T.W., Spanel, P., & Smith, D. (2011). Time-resolved selected ion flow tube mass spectrometric quantification of the volatile compounds generated by *E. coli* JM109 cultured in two different media. *Rapid Communications in Mass Spectrometry*, *25*, 2163–2172.

Costelloe, C., Metcalfe, C., Lovering, A., Mant, D., & Hay, A. D. (2010). Effect of antibiotic prescribing in primary care on antimicrobial resistance in individual patients: systematic review and meta-analysis. *BMJ*, *340*, c2096.

Delclaux, C., & Azoulay, E. (2003). Inflammatory response to infectious pulmonary injury. *European Respiratory Journal Supplement*, *42*, 10s–14s.

Dunkel, M., Schmidt, U., Struck, S., Berger, L., Gruening, B., Hossbach, J., et al. (2009). SuperScent – a database of flavors and scents. *Nucleic Acids Research*, *37*, D291–D294.

Eccles, R. (2005). Understanding the symptoms of the common cold and influenza. *Lancet Infectious Diseases*, *5*, 718–725.

Effros, R. M. (2010). Exhaled breath condensate: delusion or dilution? *Chest*, *138*, 471–472.

Enderby, B., Smith, D., Carroll, W., & Lenney, W. (2009). Hydrogen cyanide as a biomarker for *Pseudomonas aeruginosa* in the breath of children with cystic fibrosis. *Pediatric Pulmonology*, *44*, 142–147.

Fend, R., Kolk, A. H., Bessant, C., Buijtels, P., Klatser, P. R., & Woodman, A. C. (2006). Prospects for clinical application of electronic-nose technology to early detection of *Mycobacterium tuberculosis* in culture and sputum. *Journal of Clinical Microbiology*, *44*, 2039–2045.

Feng, X., & Siegel, M. M. (2007). FTICR-MS applications for the structure determination of natural products. *Analytical and Bioanalytical Chemistry*, *389*, 1341–1363.

File, T. M., Jr., Solomkin, J. S., & Cosgrove, S. E. (2011). Strategies for improving antimicrobial use and the role of antimicrobial stewardship programs. *Clinical Infectious Diseases*, *53*(Suppl. 1), S15–S22.

Fine, M. J., Smith, M. A., Carson, C. A., Mutha, S. S., Sankey, S. S., Weissfeld, L. A., et al. (1996). Prognosis and outcomes of patients with community-acquired pneumonia. A meta-analysis. *JAMA*, *275*, 134–141.

Fujitani, S., Sun, H.Y., Yu, V. L., & Weingarten, J. A. (2011). Pneumonia due to *Pseudomonas aeruginosa*: part I: epidemiology, clinical diagnosis, and source. *Chest*, *139*, 909–919.

Fu, X. A., Li, M., Biswas, S., Nantz, M. H., & Higashi, R. M. (2011). A novel microreactor approach for analysis of ketones and aldehydes in breath. *Analyst*, *136*, 4662–4666.

Geist, H. C. I. (1957). Halitosis in ancient literature. *Dental Abstracts*, *2*, 417–418.

Grosset, J. (2003). *Mycobacterium tuberculosis* in the extracellular compartment: an underestimated adversary. *Antimicrobial Agents and Chemotherapy*, *47*, 833–836.

Guaman, A.V., Carreras, A., Calvo, D., Agudo, I., Navajas, D., Pardo, A., et al. (2012). Rapid detection of sepsis in rats through volatile organic compounds in breath. *Journal of Chromatography B, Analytical Technologies in the Biomedical and Life Sciences*, *881–882*, 76–82.

Hiller, N. L., Janto, B., Hogg, J. S., Boissy, R., Yu, S., Powell, E., et al. (2007). Comparative genomic analyses of seventeen *Streptococcus pneumoniae* strains: insights into the pneumococcal supragenome. *Journal of Bacteriology*, *189*, 8186–8195.

Horvath, I., Hunt, J., Barnes, P. J., Alving, K., Antczak, A., Baraldi, E., et al. (2005). Exhaled breath condensate: methodological recommendations and unresolved questions. *European Respiratory Journal, 26*, 523–548.

Hunt, J. (2011). Condensing exhaled breath into science. *Chest, 139*, 5–6.

Ishimaru, M., Yamada, M., Nakagawa, I., & Sugano, S. (2008). Analysis of volatile metabolites from cultured bacteria by gas chromatography/atmospheric pressure chemical ionization-mass spectrometry. *Journal of Breath Research, 2*, 037021.

Jassal, M. S., Nedeltchev, G. G., Lee, J. H., Choi, S. W., Atudorei, V., Sharp, Z. D., et al. (2010). 13[C]-urea breath test as a novel point-of-care biomarker for tuberculosis treatment and diagnosis. *PLoS One, 5*, e12451.

Johansson, N., Kalin, M., Tiveljung-Lindell, A., Giske, C. G., & Hedlund, J. (2009). Etiology of community-acquired pneumonia: increased microbiological yield with new diagnostic methods. *Clinical Infectious Diseases, 50*, 202–209.

Junger, M., Vautz, W., Kuhns, M., Hofmann, L., Ulbricht, S., Baumbach, J. I., et al. (2012). Ion mobility spectrometry for microbial volatile organic compounds: a new identification tool for human pathogenic bacteria. *Applied Microbiology and Biotechnology, 93*, 2603–2614.

Kai, M., Haustein, M., Molina, F., Petri, A., Scholz, B., & Piechulla, B. (2009). Bacterial volatiles and their action potential. *Applied Microbiology and Biotechnology, 81*, 1001–1012.

Kesselmeir, A., & Staudt, M. (1999). Biogenic volatile organic compounds (VOC): an overview on emission, physiology, and ecology. *Journal of Atmospheric Sciences, 33*, 23–88.

King, J., Koc, H., Unterkofler, K., Mochalski, P., Kupferthaler, A., Teschl, G., et al. (2010). Physiological modeling of isoprene dynamics in exhaled breath. *Journal of Theoretical Biology, 267*, 626–637.

King, J., Unterkofler, K., Teschl, G., Teschl, S., Koc, H., Hinterhuber, H., et al. (2011). A mathematical model for breath gas analysis of volatile organic compounds with special emphasis on acetone. *Journal of Mathematical Biology, 63*, 959–999.

Kolk, A. H., van Berkel, J. J., Claassens, M. M., Walters, E., Kuijper, S., Dallinga, J. W., et al. (2012). Breath analysis as a potential diagnostic tool for tuberculosis. *International Journal of Tuberculosis and Lung Diseases, 16*, 777–782.

Kwak, J., & Preti, G. (2011). Volatile disease biomarkers in breath: a critique. *Current Pharmaceutical Biotechnology, 12*, 1067–1074.

Laaks, J., Jochmann, M. A., & Schmidt, T. C. (2011). Solvent-free microextraction techniques in gas chromatography. *Analytical and Bioanalytical Chemistry, 402*, 565–571.

Labows, J. N., McGinley, K. J., Webster, G. F., & Leyden, J. J. (1980). Headspace analysis of volatile metabolites of *Pseudomonas aeruginosa* and related species by gas chromatography-mass spectrometry. *Journal of Clinical Microbiology, 12*, 521–526.

Lai, S. Y., Deffenderfer, O. F., Hanson, W., Phillips, M. P., & Thaler, E. R. (2002). Identification of upper respiratory bacterial pathogens with the electronic nose. *Laryngoscope, 112*, 975–979.

Lechner, M., Karlseder, A., Niederseer, D., Lirk, P., Neher, A., Rieder, J., et al. (2005). *H. pylori* infection increases levels of exhaled nitrate. *Helicobacter, 10*, 385–390.

Li, M., Biswas, S., Nantz, M. H., Higashi, R. M., & Fu, X. A. (2011). Preconcentration and analysis of trace volatile carbonyl compounds. *Analytical Chemistry, 84*, 1288–1293.

Lim, W. S., Macfarlane, J. T., Boswell, T. C., Harrison, T. G., Rose, D., Leinonen, M., et al. (2001). Study of community-acquired pneumonia aetiology (SCAPA) in adults admitted to hospital: implications for management guidelines. *Thorax, 56*, 296–301.

Mann, F. M., Xu, M., Chen, X., Fulton, D. B., Russell, D. G., & Peters, R. J. (2009). Edaxadiene: a new bioactive diterpene from *Mycobacterium tuberculosis. Journal of American Chemical Society, 131*, 17526–17527.

Mashir, A., Paschke, K. M., van Duin, D., Shrestha, N. K., Laskowski, D., Storer, M. K., et al. (2011). Effect of the influenza A (H1N1) live attenuated intranasal vaccine on nitric oxide (FE(NO)) and other volatiles in exhaled breath. *Journal of Breath Research, 5*, 037107.

Mgode, G. F., Weetjens, B. J., Cox, C., Jubitana, M., Machang'u, R. S., Lazar, D., et al. (2011). Ability of Cricetomys rats to detect *Mycobacterium tuberculosis* and discriminate it from other microorganisms. *Tuberculosis (Edinburgh)*, *92*, 182–186.

Mgode, G. F., Weetjens, B. J., Nawrath, T., Cox, C., Jubitana, M., Machang'u, R. S., et al. (2011). Diagnosis of tuberculosis by trained African giant pouched rats and confounding impact of pathogens and microflora of the respiratory tract. *Journal of Clinical Microbiology*, *50*, 274–280.

Miekisch, W., Schubert, J. K., & Noeldge-Schomburg, G. F. (2004). Diagnostic potential of breath analysis – focus on volatile organic compounds. *Clinica Chimica Acta*, *347*, 25–39.

Montuschi, P. (2007). Analysis of exhaled breath condensate in respiratory medicine: methodological aspects and potential clinical applications. *Therapeutic Advances in Respiratory Diseases*, *1*, 5–23.

Nawrath, T., Mgode, G. F., Weetjens, B., Kaufmann, S. H., & Schulz, S. (2012). The volatiles of pathogenic and nonpathogenic mycobacteria and related bacteria. *Beilstein Journal of Organic Chemistry*, *8*, 290–299.

Neuhauser, M. M., Weinstein, R. A., Rydman, R., Danziger, L. H., Karam, G., & Quinn, J. P. (2003). Antibiotic resistance among gram-negative bacilli in US intensive care units: implications for fluoroquinolone use. *JAMA*, *289*, 885–888.

O'Hara, M., & Mayhew, C. A. (2009). A preliminary comparison of volatile organic compounds in the headspace of cultures of *Staphylococcus aureus* grown in nutrient, dextrose and brain heart bovine broths measured using a proton transfer reaction mass spectrometer. *Journal of Breath Research*, *3*, 027001.

Pathak, C. M., Bhasin, D. K., & Khanduja, K. L. (2004). Urea breath test for *Helicobacter pylori* detection: present status. *Tropical Gastroenterology*, *25*, 156–161.

Pauling, L., Robinson, A. B., Teranishi, R., & Cary, P. (1971). Quantitative analysis of urine vapor and breath by gas-liquid partition chromatography. *Proceedings of the National Academy of Sciences of the United States of America*, *68*, 2374–2376.

Phillips, M. (1997). Method for the collection and assay of volatile organic compounds in breath. *Analytical Biochemistry*, *247*, 272–278.

Phillips, M., Basa-Dalay, V., Blais, J., Bothamley, G., Chaturvedi, A., Modi, K. D., et al. (2012). Point-of-care breath test for biomarkers of active pulmonary tuberculosis. *Tuberculosis (Edinburgh)*, *92*, 314–320.

Phillips, M., Basa-Dalay, V., Bothamley, G., Cataneo, R. N., Lam, P. K., Natividad, M. P., et al. (2010). Breath biomarkers of active pulmonary tuberculosis. *Tuberculosis (Edinburgh)*, *90*, 145–151.

Phillips, M., Cataneo, R. N., Condos, R., Ring Erickson, G. A., Greenberg, J., La Bombardi, V., et al. (2007). Volatile biomarkers of pulmonary tuberculosis in the breath. *Tuberculosis (Edinburgh)*, *87*, 44–52.

Phillips, M., & Greenberg, J. (1992). Ion-trap detection of volatile organic compounds in alveolar breath. *Clinical Chemistry*, *38*, 60–65.

Phillips, M., Greenberg, J., & Awad, J. (1994). Metabolic and environmental origins of volatile organic compounds in breath. *Journal of Clinical Pathology*, *47*, 1052–1053.

Prach, L., Kirby, J., Keasling, J. D., & Alber, T. (2010). Diterpene production in Mycobacterium tuberculosis. *FEBS Journal*, *277*, 3588–3595.

Preti, G., Thaler, E., Hanson, C. W., Troy, M., Eades, J., & Gelperin, A. (2009). Volatile compounds characteristic of sinus-related bacteria and infected sinus mucus: analysis by solid-phase microextraction and gas chromatography-mass spectrometry. *Journal of Chromatography B, Analytical Technologies in the Biomedical and Life Sciences*, *877*, 2011–2018.

Purkhart, R., Kohler, H., Liebler-Tenorio, E., Meyer, M., Becher, G., Kikowatz, A., et al. (2011). Chronic intestinal Mycobacteria infection: discrimination via VOC analysis in exhaled breath and headspace of feces using differential ion mobility spectrometry. *Journal of Breath Research*, *5*, 027103.

Ramirez, J. A., & Anzueto, A. R. (2011). Changing needs of community-acquired pneumonia. *Journal of Antimicrobial Chemotherapy, 66*(Suppl. 3), iii3–iii9.

Robroeks, C. M., van Berkel, J. J., Dallinga, J. W., Jobsis, Q., Zimmermann, L. J., Hendriks, H. J., et al. (2010). Metabolomics of volatile organic compounds in cystic fibrosis patients and controls. *Pediatric Research, 68*, 75–80.

Rosias, P. (2012). Methodological aspects of exhaled breath condensate collection and analysis. *Journal of Breath Research, 6*, 027102.

Rosias, P., Robroeks, C., Hendriks, J., Dompeling, E., & Jobsis, Q. (2004). Exhaled breath condensate: a space odessey, where no one has gone before. *European Respiratory Journal, 24*, 189–190. author reply 190.

Saad, R. J., & Chey, W. D. (2007). Breath tests for gastrointestinal disease: the real deal or just a lot of hot air? *Gastroenterology, 133*, 1763–1766.

Savelev, S. U., Perry, J. D., Bourke, S. J., Jary, H., Taylor, R., Fisher, A. J., et al. (2011). Volatile biomarkers of *Pseudomonas aeruginosa* in cystic fibrosis and noncystic fibrosis bronchiectasis. *Letters in Applied Microbiology, 52*, 610–613.

Schulz, S., & Dickschat, J. S. (2007). Bacterial volatiles: the smell of small organisms. *Natural Product Reports, 24*, 814–842.

Scott-Thomas, A., Pearson, J., & Chambers, S. (2011). Potential sources of 2-aminoaceto-phenone to confound the *Pseudomonas aeruginosa* breath test, including analysis of a food challenge study. *Journal of Breath Research, 5*, 046002.

Scott-Thomas, A. J., Syhre, M., Pattemore, P. K., Epton, M., Laing, R., Pearson, J., et al. (2010). 2-Aminoacetophenone as a potential breath biomarker for *Pseudomonas aeruginosa* in the cystic fibrosis lung. *BMC Pulmonary Medicine, 10*, 56.

Scotter, J. M., Allardyce, R. A., Langford, V. S., Hill, A., & Murdoch, D. R. (2006). The rapid evaluation of bacterial growth in blood cultures by selected ion flow tube–mass spectrometry (SIFT-MS) and comparison with the BacT/ALERT automated blood culture system. *Journal of Microbiological Methods, 65*, 628–631.

Shestivska, V., Nemec, A., Drevinek, P., Sovova, K., Dryahina, K., & Spanel, P. (2011). Quantification of methyl thiocyanate in the headspace of *Pseudomonas aeruginosa* cultures and in the breath of cystic fibrosis patients by selected ion flow tube mass spectrometry. *Rapid Communications in Mass Spectrometry, 25*, 2459–2467.

Skogerson, K., Wohlgemuth, G., Barupal, D. K., & Fiehn, O. (2011). The volatile compound BinBase mass spectral database. *BMC Bioinformatics, 12*, 321.

Smith, T. (2011). Breath analysis: clinical research to the end-user market. *Journal of Breath Research, 5*, 032001.

Spanel, P., & Smith, D. (2011). Comment on 'influences of mixed expiratory sampling parameters on exhaled volatile organic compound concentrations'. *Journal of Breath Research, 5*, 048001.

Strand, N., Bhushan, A., Schivo, M., Kenyon, N. J., & Davis, C. E. (2010). Chemically polymerized polypyrrole for on-chip concentration of volatile breath metabolites. *Sensors and Actuators B: Chemical, 143*, 516–523.

Suckling, D. M., & Sagar, R. L. (2011). Honeybees *Apis mellifera* can detect the scent of *Mycobacterium tuberculosis*. *Tuberculosis (Edinburgh), 91*, 327–328.

Syhre, M., & Chambers, S. T. (2008). The scent of *Mycobacterium tuberculosis*. *Tuberculosis (Edinburgh), 88*, 317–323.

Syhre, M., Manning, L., Phuanukoonnon, S., Harino, P., & Chambers, S. T. (2009). The scent of *Mycobacterium tuberculosis* – part II breath. *Tuberculosis (Edinburgh), 89*, 263–266.

Teranishi, R., Mon, T. R., Robinson, A. B., Cary, P., & Pauling, L. (1972). Gas chromatography of volatiles from breath and urine. *Analytical Chemistry, 44*, 18–20.

Thekedar, B., Oeh, U., Szymczak, W., Hoeschen, C., & Paretzke, H. G. (2011). Influences of mixed expiratory sampling parameters on exhaled volatile organic compound concentrations. *Journal of Breath Research, 5*, 016001.

Thorn, R. M., & Greenman, J. (2012). Microbial volatile compounds in health and disease conditions. *Journal of Breath Research, 6*, 024001.

Thorn, R. M., Reynolds, D. M., & Greenman, J. (2010). Multivariate analysis of bacterial volatile compound profiles for discrimination between selected species and strains in vitro. *Journal of Microbiological Methods, 84*, 258–264.

Tonkin-Crine, S., Yardley, L., & Little, P. (2011). Antibiotic prescribing for acute respiratory tract infections in primary care: a systematic review and meta-ethnography. *Journal of Antimicrobial Chemotherapy, 66*, 2215–2223.

Trivedi, A., Singh, N., Bhat, S. A., Gupta, P., & Kumar, A. (2012). Redox biology of tuberculosis pathogenesis. *Advances in Microbial Physiology, 60*, 263–324.

Ustianowski, A. (2012). Diagnostics for community-acquired and atypical pneumonia. *Current Opinion in Pulmonary Medicine, 18*, 259–263.

Weetjens, B. J., Mgode, G. F., Machang'u, R. S., Kazwala, R., Mfinanga, G., Lwilla, F., et al. (2009). African pouched rats for the detection of pulmonary tuberculosis in sputum samples. *International Journal of Tuberculosis and Lung Diseases, 13*, 737–743.

Whittle, C. L., Fakharzadeh, S., Eades, J., & Preti, G. (2007). Human breath odors and their use in diagnosis. *Annals of the New York Academy of Sciences, 1098*, 252–266.

WHO, 2010. Global Tuberculosis Control.

Woolfenden, E. (2010). Sorbent-based sampling methods for volatile and semi-volatile organic compounds in air. Part 2. Sorbent selection and other aspects of optimizing air monitoring methods. *Journal of Chromatography A, 1217*, 2685–2694.

Zechman, J. M., Aldinger, S., & Labows, J. N., Jr. (1986). Characterization of pathogenic bacteria by automated headspace concentration-gas chromatography. *Journal of Chromatography, 377*, 49–57.

Zhang, J., Fang, A., Wang, B., Kim, S. H., Bogdanov, B., Zhou, Z., et al. (2011). iMatch: a retention index tool for analysis of gas chromatography-mass spectrometry data. *Journal of Chromatography A, 1218*, 6522–6530.

CHAPTER THREE

Polymicrobial Multi-functional Approach for Enhancement of Crop Productivity

Chilekampalli A. Reddy[1], Ramu S. Saravanan

Department of Microbiology and Molecular Genetics, Michigan State University, East Lansing, MI, USA
[1]Corresponding author: E-mail: reddy@msu.edu

Contents

Abstract

There is an increasing global need for enhancing the food production to meet the needs of the fast-growing human population. Traditional approach to increasing agricultural productivity through high inputs of chemical nitrogen and phosphate fertilizers and pesticides is not sustainable because of high costs and concerns about global warming, environmental pollution, and safety concerns. Therefore, the use of naturally occurring soil microbes for increasing productivity of food crops is an attractive eco-friendly, cost-effective, and sustainable alternative to the use of chemical fertilizers

Advances in Applied Microbiology, Volume 82
ISSN 0065-2164, http://dx.doi.org/10.1016/B978-0-12-407679-2.00003-X

and pesticides. There is a vast body of published literature on microbial symbiotic and nonsymbiotic nitrogen fixation, multiple beneficial mechanisms used by plant growth-promoting rhizobacteria (PGPR), the nature and significance of mycorrhiza–plant symbiosis, and the growing technology on production of efficacious microbial inoculants. These areas are briefly reviewed here. The construction of an inoculant with a consortium of microbes with multiple beneficial functions such as N_2 fixation, biocontrol, phosphate solubilization, and other plant growth-promoting properties is a positive new development in this area in that a single inoculant can be used effectively for increasing the productivity of a broad spectrum of crops including legumes, cereals, vegetables, and grasses. Such a polymicrobial inoculant containing several microorganisms for each major function involved in promoting the plant growth and productivity gives it greater stability and wider applications for a range of major crops. Intensifying research in this area leading to further advances in our understanding of biochemical/molecular mechanisms involved in plant–microbe–soil interactions coupled with rapid advances in the genomics–proteomics of beneficial microbes should lead to the design and development of inoculants with greater efficacy for increasing the productivity of a wide range of crops.

1. INTRODUCTION

To meet the ever-increasing needs for food by the fast-growing human population in the world, currently believed to be a little over seven billions and estimated to reach ~10 billion by 2050, is one of the greatest global challenges of the 21st millennium (UN World Population Prospects: http://esa.un.org/wpp/Other-Information/faq.htm). Also, as per UN statistics, approximately one billion people (http://www.wfp.org/hunger/stats?gclid=CPXYyLuCjbECFeUBQAod7FamDA) in the world suffer from hunger. It has further been reported that the world food production has to increase by at least 70% by the year 2050 to meet the human food needs as per FAO of the UN (http://www.fao.org/f1leadmin/templates/wsfs/docs/expert_paper/How_to_Feed_the_World_in_2050.pdf). This increased need for food would impose a large stress on agriculture production worldwide. Meeting these expected needs for food is rather difficult because of the steep rise in the prices of energy and food grains worldwide, growing concern over global warming, continuing shrinkage in arable land, and increasing awareness of the need to protect our environment from the standpoint of global warming, and preservation of human and animal health. Using conventional agricultural approaches, high levels of nitrogen fertilizers (such as urea and ammonium nitrate) are used to increase food production. In fact, the input of nitrogenous fertilizers constitutes the single highest nonfarm cost in raising many food crops and represents about 50% of the operational costs in agriculture (Dodd & Ruiz-Lozano, 2012; Gutierrez, 2012; Maheshwari, 2011; Serraj, 2004). However, of the estimated 120 Tg/year of N used for global food production, only about 10% is actually consumed by the people (Robertson & Vitousek, 2009). Also, manufacture of nitrogen fertilizers requires fossil fuel energy sources, which are in short supply, are expensive, contribute adversely to global warming, affect environmental quality, and constitute health hazard to humans (Altman & Hasegawa, 2012; Gepts, 2012). Furthermore, 50–60% of the N fertilizers applied to the field is lost by leaching into the soil or released into the environment as N gases (such as NO and N_2O), which contribute to global warming (Gutierrez, 2012). Excess nitrogen runoff leads to eutrophication of terrestrial and aquatic systems. Today, the size of the hypoxic zone in northern Gulf of Mexico is >20,000 km^2, and there is good evidence that eutrophication (over-enrichment of the Gulf waters that drives the formation of the hypoxic zone) is caused by reactive N rather than P or other nutrients (Robertson & Vitousek, 2009). Therefore, there is a vital need to develop innovative, effective, and affordable approaches to increase crop productivity on a long-term sustainable basis in energy efficient and eco-friendly manner. Developing microbial inoculants that contain N_2-fixing bacteria, which reduce or eliminate the need for added nitrogen fertilizer for a variety of important leguminous and cereal crops, would contribute to a substantial reduction in the cost of food production, positively impact the world economy and reduce potential health and environmental hazards associated with the heavy use of nitrogenous fertilizers (Altman & Hasegawa, 2012; Dodd & Ruiz-Lozano, 2012; Miransari, 2011; Wang, Yang, Tang, & Zhu, 2012).

Phosphorus (P) is perhaps next only to nitrogen (N) in terms of its importance as an essential nutrient for optimal productivity of various food crops (Babalola, 2010; Diallo et al., 2011; Siddiqui, 2006). Heavy use of P-fertilizers, in addition to their significant cost to the farmer, is known to result in significant environmental costs because P runoff into aquatic systems leads to eutrophication lakes and other aquatic habitats (Lee & Jones, 1986; Sharpley et al., 2003). A number of rhizosphere microbes have been shown to be quite efficient in solubilizing insoluble phosphates (that are not accessible to the plant) into a form of P that is utilizable by the plant for its metabolic needs. Hence, P-solubilizing soil microbes have the potential to increase crop yields when used alone or in combination with other plant growth-promoting microorganisms (see Section 3).

Different classes of chemical pesticides are used rather extensively world-wide for controlling important diseases of food crops and increase yields (Sharma, 2009; Varma, 2008). However, chemical pesticides are a significant cost item to the farmers. Furthermore, a number of pesticides persist in the environment and are toxic to humans and animals (even at parts per billion levels in some cases). Therefore, there have been several studies in identifying soil bacteria and fungi that can serve as biocontrol agents and eliminate or reduce the need for pesticides (Hayat, Ali, Amara, Khalid, & Ahmed, 2010; Lugtenberg & Kamilova, 2009; Nihorimbere et al., 2012; Vincent, Goettel, & Lazarovits, 2007). Several species of *Trichoderma, Bacillus, Pseudomonas* and of mycorrhiza are common biocontrol agents that have been studied in considerable detail (Babalola, 2010; Johansson, Paul, & Finlay, 2004; Paulitz & Belanger, 2001; Siddiqui, 2006). Mechanisms for biocontrol employed by different microbes may include one or more of the mechanisms such as antibiosis, antagonism, competitive exclusion, and induction of innate systemic resistance (see Section 3); these and additional mechanisms involved in biocontrol have been reviewed (Dessaux, Hinsinger, & Lemanceau, 2010; Dodd & Ruiz-Lozano, 2012; Koltai & Kapulnik, 2010; Maheshwari, 2011; Miransari, 2011; Varma & Kharkwal, 2009).

Soil is a rich storehouse of diverse communities of microbes with multifaceted metabolic activities. A number of diverse phylogenetic groups of microbes in nature have been described, which increase crop productivity by various plant growth-promoting mechanisms (Dutta & Podile, 2010; Hayat et al., 2010). Numerous investigators have developed microbial inoculants that confer beneficial properties, such as nitrogen fixation, uptake of P and other mineral nutrients, biocontrol, and plant hormone production, for increasing the crop productivity (Bashan, Holguin, & de-Bashan, 2004; Dodd & Ruiz-Lozano, 2012; Miransari, 2011). Development of stable, efficacious, and eco-friendly microbial formulations that contain phylogenetically diverse and naturally occurring soil microbes with multiple complementary functions designed to enhance the productivity of a broad spectrum of crops with reduced input of N and P fertilizers and pesticides is the ideal goal. It is encouraging that some real progress is being made in this direction, and perhaps better microbial formulations for increasing crop productivity may become available soon.

We have presented here a relatively brief but representative coverage of recent information on diazotrophs, plant growth-promoting microorganisms, mycorrhizal symbioses, and microbial inoculants, including some recent research on polymicrobial formulations. A number of excellent books and reviews have been published on the topical areas included in this chapter (Altman & Hasegawa, 2012; Arora, Khare, & Maheshwari, 2011; Bending, Aspray, & Whipps, 2006; Bonfante & Anca, 2009; Cooper & Scherer, 2012; Dodd & Ruiz-Lozano, 2012; Fulton, 2011; Hayat et al., 2010; Khan et al., 2010; Maheshwari, 2011; Malusa, Sas-Paszt, & Ciesielska, 2012; Miransari, 2011; Parniske, 2008; Siddiqui, 2006; Sturmer, 2012; Wang et al., 2012; Zamioudis & Pieterse, 2012). Because of space limitations, we have primarily referred to the relevant recent publications and reviews so that these will help the reader to access original publications.

2. NITROGEN-FIXING RHIZOBACTERIA

2.1. Symbiotic Nitrogen Fixers

Nitrogen (N_2) is essential for cellular synthesis of enzymes, proteins, chlorophyll, DNA, and RNA and is the single-most important essential nutrient that positively affects plant growth and productivity. Even though 79% of the atmospheric air contains N_2, the plant is unable to use it as it needs reduced form of N_2 such as ammonia. Plants also can use NO_3, NH_2, and a few other forms of nitrogen. Nitrogen needs of the crops are provided abiologically by the application of chemical fertilizers, a practice that is not sustainable, expensive, and polluting. Nitrogen is provided biologically by specialized microorganisms that have the unique ability to convert atmospheric N_2 to ammonia, mediated by the enzyme nitrogenase. Microbial N_2-fixation is probably the most important biologically mediated process on earth, after photosynthesis. Biological N_2-fixation (BNF) is an eco-friendly, cost-effective, and complementary process to reduce the usage of commercial chemical fertilizers or in some cases as an alternative to the use of nitrogen fertilizers. Approximately, 65%–70% of the nitrogen currently utilized in agriculture is actually derived from BNF (about 180 million metric tons per annum globally) and will continue to be an important source of nitrogen in future sustainable crop production systems (Cooper & Scherer, 2012; Dixon & Kahn, 2004; Pawlowska & Taylor, 2004; Serraj, 2004). BNF is one of the most important factors in raising global food production in a sustainable manner. In the case of BNF, the efficiency of nitrogen transfer to the plant is quite high (~95%) because nitrogen supplied by BNF is directly delivered to the plant for its use. The high efficiency of BNF is also due to the fact that there is a close two-way communication between the plant and the N_2-fixers so that N_2-fixation by the later is modulated (up or down) to fit the needs of the plant. In the case of chemical fertilizer on the other hand, nearly 50% of the fertilizer applied to the soil is lost before it ever reaches the plant. This loss of fertilizer is not only a loss to the farmer but also causes multifold harm to the environment and to public health. Therefore, nitrogen fertilizer usage is not cost-effective, adversely impacts soil health, pollutes the environment, and is not sustainable in the long run. Agricultural practices that enhance BNF would reduce input of chemical fertilizers and favorably impact the global agricultural economy and environmental health. BNF has been studied extensively and several recent books and reviews focus on this topic (Cooper, 2004; Cooper & Scherer, 2012; Dixon & Kahn, 2004; Emerich & Krishnan, 2009; Serraj, 2004).

In the BNF process, fixation of N_2 in air to ammonia (NH_3) is represented by the equation:

$$N_2 + 8\,H^+ + 8\,e^- \rightarrow 2\,NH_3 + H_2$$

This is an energy-intensive process requiring 16 equivalents of ATP and production of 2 NH_3 and one H_2. Soil N_2-fixing bacteria (often called 'diazotrophs') distributed worldwide in nature are literally "nitrogen nano factories", and play the central role in BNF. There is a wide phylogenic diversity in N_2-fixing bacteria. For example, bacteria that fix N_2 in plant root nodules include 12 genera and >50 species and include both α-proteobacteria and β-proteobacteria. Also, there is substantial variation among N_2-fixing bacteria in terms of their relative efficiency in fixing N_2. New genomic information indicates that 5% of all prokaryotes carry N_2 fixation genes. Relationship between diazotrophs and plants range from rather loose associations of heterotrophic bacteria around plant roots, endophytic bacteria residing in the vascular tissues of tropical grasses, and symbiotic N_2-fixing bacteria, that carryout highly evolved and complex symbiosis involving morphological differentiation of both microbe and plant in specialized root structures called nodules (Dixon & Kahn, 2004; Emerich & Krishnan, 2009; Shtark et al., 2011; Toro, Azcon, & Barea, 1997). Symbiotic N_2-fixers represented by Rhizobiales can fix N_2 only in symbiotic association with legume root nodules. It has been estimated that symbiotic N_2-fixation (SNF) in legumes in the U.S. in 2001 was equivalent to the use of approximately 17 million tons of atmospheric N_2, worth about 8 billion U.S. dollars in fertilizer value. Free-living N_2-fixers such as *Azospirillum* are closely associated with the root systems of the host plant (associative N_2-fixers), while other N_2-fixers such as *Azotobacter* fix N_2 in a free-living state in the soil. Frankia, an actinomycete, is involved in SNF with a range of angiosperms such as *Alnus* and *Casuarina* (Wheeler & Miller, 1990).

Rhizobiales including the genera *Rhizobium, Bradyrhizobium, Mesorhizobium, Sinorhizobium Azorhizobium* and *Allorhizobium* are the major players in symbiotic N_2-fixation in nature. *Methylobacterium,* also included in this group, appears to be of lesser importance. Legumes (as all plants) are unable to use N_2 in the air as a nitrogen source for various cellular processes. Symbiotic N_2-fixing bacteria in the root nodules of selected plants such as legumes fix N_2 into NH_3, which is usable by the plant. Legume–Rhizobia symbiosis involves the provision of photosynthate rich in organic carbon compounds to the bacteria, and the bacteria in turn reduce atmospheric N_2 to ammonia and make it available for the plant. Important biochemical reactions involved in SNF occur mainly through symbiotic association of N_2-fixing rhizobia in legumes resulting in nitrogenase-mediated reduction of N_2 into NH_3, which can be utilized by the plant (Baldani & Baldani, 2005; Emerich & Krishnan, 2009). The rhizobial infection begins when the bacteria enter into roots in a host-controlled manner (Limpens & Bisseling, 2003) and form intimate symbiotic relationships with legumes by producing N_2-fixing root nodules in response to flavonoid molecules released as signals by the legume host (Cooper, 2004; Cooper & Scherer, 2012; Emerich & Krishnan, 2009; Long, 1989). These plant signals induce the expression of nodulation genes in rhizobia, which in turn produce lipochitooligosaccharide (LCO) signals that trigger mitotic cell division in roots, leading to nodule formation (Deakin & Broughton, 2009; Dénarié & Cullimore, 1993; Matiru & Dakora, 2005; Ortiz-Castro, Contreras-Cornejo, Macias-Rodriguez, & López-Bucio, 2009). Hydrogen is generated as a product of the nitrogenase reaction. Uptake hydrogenase takes up this hydrogen, which is then oxidized through an electron transport chain-liberating ATP, as an energy source. This ATP is available to drive various metabolic reactions.

There are a number of factors that affect the nodulation on legume roots including host-microbe symbiotic compatibility, physicochemical conditions of the soil, and presence of both known and unknown biomolecules such as flavonoids, lumichrome, polysaccharides, and hormones (Matiru & Dakora, 2005; Mortier, Holsters, & Goormachtig, 2012). It is a molecular dialog between the host plant and a compatible strain of *Rhizobium,* which serves to initiate the development of the root nodules, the site of SNF (Cooper, 2007; Cooper & Scherer, 2012; Deakin & Broughton, 2009). Rapidly developing genomics and proteomics technology opens new vistas in obtaining a better understanding of BNF and in developing new biotechnological approaches to improve the efficiency of SNF in the production of major food and vegetable crops.

2.2. Nonsymbiotic Nitrogen Fixers

The rhizosphere is generally described as the soil immediately surrounding the plant root and is under the influence of the root (Morgan, Bending, & White, 2005; Nihorimbere, Ongena, Smargiassi, & Thonart, 2011). It actually consists of three distinct components: rhizosphere per se, rhizoplane (the plant root surface), and the root itself (see Fig. 3.1). Nourished by the carbon-rich plant photosynthate, an abundant and highly complex microbial flora (approximately 10^7 to 10^9 microorganisms/g soil) reside in the rhizosphere (Raab & Lipson, 2010). A large majority of

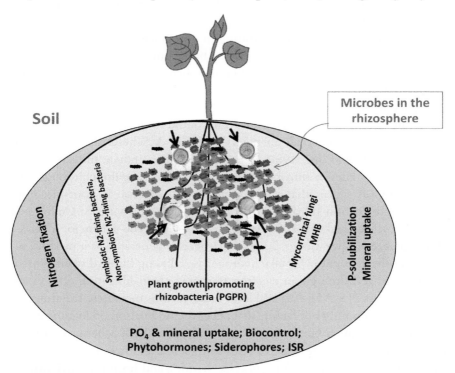

Figure 3.1 *Schematic diagram of dynamic interaction between plant, microbes, and soil.* The inner pink oval presents three major microbial groups in the rhizosphere: symbiotic and nonsymbiotic N_2-fixing bacteria, plant growth-promoting rhizobacteria (PGPR), and mycorrhizal fungi [including mycorrhizal helper bacteria (MHB)]. In the outer blue oval, the respective beneficial functions provided by these three groups of microorganisms to the plant are shown. MHB facilitate mycorrhizal colonization of the plant. Diverse rhizosphere microbes are represented by the colored structures and the round bodies in the figure (arrows) represent mycorrhizal spores. (For interpretation of the references to color in this figure legend, the reader is referred to the online version of this book.)

these are bacteria and fungi. Compared to the bulk soil, rhizosphere contains 2–20 times higher number of bacteria and 10–20 times higher number of fungi. Arbuscular mycorrhizal fungi (described in Section 4) account for about 25% of the total microbial mass and about 80% of the fungal biomass in the rhizosphere. The rhizodeposition, one of the main factors influencing the rhizosphere, refers to the organic or inorganic nutrients of the root within the soil (Emerich & Krishnan, 2009; Nihorimbere et al., 2009; Zhang et al., 2012). As much as 15–40% of the total photosynthetic carbon of the plant is secreted into the rhizosphere and is the most important energy resource for the complex microbial flora in the rhizosphere environment (Dixon & Kahn, 2004); however, these expenses are amply compensated by the important benefits conferred to the plant by bacteria and fungi in the rhizosphere. Roots also influence the rhizosphere by creating a negative oxygen gradient (root respiration) by absorbing water and mineral salts.

Many rhizosphere microbes live as endophytes in the host plants. Endophytes are bacteria or fungi that live within in a plant (for at least part of their life) within intercellular spaces, tissue cavities, or vascular bundles without harming the host and often benefit the host. Endophytes (also called endosymbionts) are isolated from monocots and dicots and in fact in every plant species that has been examined to date. A number of diazotrophic bacteria such as *Azospirillum,* which reduce N_2 to ammonia and make it available to the plant, are endophytes (Cooper & Scherer, 2012; Emerich & Krishnan, 2009). Endophytes have also been shown to offer protection against certain plant pathogens and improve plant's ability to cope with drought and other abiotic stresses (see below). Beneficial endophytes have been described in many commercially important cereals such as corn, maize, sorghum, wheat, and rice as well as in other crops such as tomato, potato, carrots, soybean, and sugarcane. When plants lack essential mineral elements such as P or N, symbiotic relationships with the rhizosphere microbes can be quite beneficial in promoting the plant growth (Morgan et al., 2005; Zamioudis & Pieterse, 2012).

A range of plant growth-promoting rhizobacteria (PGPR) that associate with non-leguminous plants and fix atmospheric N_2 to NH_3 to meet the N needs of the plant are called nonsymbiotic N_2-fixers. Nonsymbiotic N_2-fixers are of two types: free-living soil bacteria and associative bacteria (Hayat et al., 2010). Free-living nitrogen-fixing bacteria and associative N_2-fixers are exemplified by *Azotobacter* and *Azospirillum* species, respectively. Neither group forms N_2-fixing root nodules. Azospirillum and *Pseudomonas* species efficiently attach to the root surfaces of non-legumes and enhance crop yields (Gupta, Gopal, & Tilak, 2000; Herschkovitz et al., 2005; Lugtenberg

& Kamilova, 2009; Miransari, 2011; Somers, Vanderleyden, & Srinivasan, 2004). Some of these organisms are endophytic and are associated with the root of the host (associative N_2-fixers). These associative N_2-fixers participate in interaction with C3 and C4 plants (e.g. rice, wheat, maize, sugarcane, and cotton), and significantly increase their vegetative growth and yield (Hayat et al., 2010). Azotobacter species (*Azotobacter vinelandii* and *Azotobacter chroococcum*) are free-living heterotrophic diazotrophs that depend on an adequate supply of reduced C compounds such as sugars in the plant photosynthate for their energy source (Dilworth & Glenn, 1984; Matiru & Dakora, 2005; Promé, 1996). Several genera of *Enterobacteriaceae* include diazotrophs, particularly those isolated from the rhizosphere of rice as exemplified by *Klebsiella pneumonia* and *Enterobacter cloacae*. *Herbaspirillum* is an endophyte, which colonizes sugarcane, rice, maize, sorghum, and other cereals (Gutierrez & Romero, 2001; Raab & Lipson, 2010).

1. *Azospirillum.* Genus *Azospirillum* contains seven species that are widely distributed. Azospirillum species are microaerophilic, nonsymbiotic, diazotrophic bacteria (associative N_2-fixers) that display a number of PGPR activities (see Section 3). *Azospirillum,* a well-recognized associative diazotroph, is able to promote growth and yield of host plants (Bashan et al., 2004; Fasciglione, Casanovas, Yommi, Sueldo, & Barassi, 2012). *Azospirillum* species colonize roots of rice, maize, wheat, sorghum (and other cereals), sugarcane, cotton, grasses, and vegetables such as tomato, pepper, beets, sweet potato, and eggplant. When inoculated onto cereal roots, *Azospirillum* strains multiply and form small aggregates, mainly in zones of root elongation and root hair. These organisms live in the intercellular spaces of the plant's vascular system, take dissolved N_2 from the sap flow, and convert it into ammonia nitrogen and amines for use by the plants (see Bashan et al., 2004). *Azospirillum* also grows extensively in the rhizosphere of graminaceous plants (without forming nodules unlike rhizobia) increasing yield because of nonSNF as well as multiple plant growth promotion (PGP) properties (Kennedy & Tchan, 1992). *Azospirillium lipoferum* isolates constituted 85% of the *Azospirillium* isolated in rice soils (James, Gyaneshwar, Barraquio, Mathan, & Ladha, 2000). In another study, *A. lipoferum* and *A. brasilense* significantly increased the sugarcane yield when inoculated into sugarcane rhizosphere and they were recovered from roots, stems, and leaves. Both *A. brasilense* and *A. lipoferum* are found in roots, stems, and leaves of the sugarcane plant, while *Azospirillum amazonense* is found in roots and stems (see Bashan et al., 2004). Some isolates of *Azospirillum* from endorhizosphere of

sugarcane (*Saccharum* sp.) and rye grass (*Lolium perenne*) exhibited high nitrogenase activity compared to a number of other isolates of *Azospirillum,* indicating considerable strain variation (Gangwar & Kaur, 2009). Ability of *Azospirillum* to increase growth and yield of host plants has been attributed less to its ability to fix N_2 and more to its ability to increase growth of root hairs and the number of lateral roots on cotton plants resulting in a better uptake of nutrients by the host plant (Bashan, 1998). Wheat plants inoculated with *Azospirillium* increased growth and grain yield in greenhouse trials. [15]N tracer studies showed that only 7–12% of N in the wheat plant was derived from nonSNF. The rest of the grain yield increase was attributed to the PGP activities of *Azospirillium*. *Azospirillium brasilense* inoculation increased chlorophyll concentrations of leaves and vegetative development of grapevine rootstocks (Fibach-Paldi, Burdman, & Okon, 2012; Sabir et al., 2012).

2. *Azotobacter.* Genus *Azotobacter* is an aerobic, free-living, nonsymbiotic N_2-fixer that is often found in the root tissue. It contains six species of which *A. vinelandii* and *A. chroococcum* are better known. Inoculation of wheat with *Azotobacter* replaced up to 50% of the urea–N in greenhouse trials under aseptic (gnotobiotic) conditions (Hegazi et al., 1998; Soliman, Seeda, Aly, & Gadalla, 1995). Inoculation with *Azotobacter* spp. can increase cotton yield by 15–28% (Iruthayaraj, 1981) as a result of BNF, production of antibacterial and antifungal compounds, growth regulators, and siderophores (Pandey & Kumar, 1989). There is a general consensus that increased yields in *Azotobacter*-inoculated soils are primarily due to the PGPR properties of these bacteria.

3. *Azorhizobium.* *Azorhizobium* caulinodans-inoculated wheat plants had increased dry weight and N content (Matthews, Sparkes, & Bullard, 2001). When N_2-fixing (nif1) and non-N_2-fixing (nif2) strains of *A. caulinodans* were used as inoculants, increased growth of wheat was observed in plants inoculated with either nif1 or nif2, as compared to the uninoculated plants, suggesting that the beneficial effect on wheat yields was from the production of PGP substances rather than by BNF. Further studies showed that using *A. caulinodans* inoculant can save up to 50% of N input both under greenhouse and field conditions (Saleh, Mekhemar, El-Soud, Ragab, & Mikhaeel, 2001).

4. *Azoarcus.* *Azoarcus* are gram-negative, endophytic, nonsymbiotic N_2-fixers originally isolated from Kallar grasses (Reinhold-Hurek et al., 1993). *Azoarcus* sp. colonized rice roots and showed both intracellular and intercellular growth. Nitrogen fixation was quite efficient in that nitrogenase activity was an order of magnitude higher than that seen in

bacteriods of symbiotic diazotrophs. However, practical significance of *Azoarcus* for enhancing crop production is not clear.

5. *Burkholderia.* Genus *Burkholderia,* belongs to β-proteobacteria, and contains 29 species of which some are non-symbiotic N_2-fixers and also have PGP properties (Vandamme, Goris, Chen, Vos de, & Willems, 2002). There is considerable variation among species of *Burkholderia* in terms of their growth-promoting and N_2-fixing abilities. *B. vietnamiensis* was isolated from rice roots and adhering soil, as also from the interior of roots, stems, and leaves of rice (Baldani, Baldani, & Dobereiner, 2000). It satisfies about 19–31% of the N needs of the rice plant, thus substantially lowering the cost of N-fertilizers. It increased rice plant biomass by up to 69% under gnotobiotic conditions (Baldani, Baldani, & Dobereiner, 2000). Some strains of *Burkholderia,* (similar to rhizobia), form symbiotic N_2-fixing root nodules in legumes of genus *Mimosa* (Gyaneshwar et al., 2011). Some species of genus *Burkholderia,* such as *B. phymatum,* are quite promiscuous and nodulate several important legumes including common bean (*Phaseolus vulgaris*). *B. tropicalis* is antagonistic to plant nematodes and may have a place in some of the bioinoculants considered for field trials (Meyer, Massoud, Chitwood, & Roberts, 2000). Consistent with the reported species variation in Genus *Burkholderia,* it has been reported that some species of this genus cause grain and seedling rot of rice (Nakata, 2002). Furthermore, several species of *Burkholderia* are important causes of diseases in human and animals, although it appears that these pathogenic strains may actually belong to a new genus based on molecular evidence.

6. *Enterobacteriaceae.* Several species of *Enterobacteriaceae* were isolated and identified from rice fields, and they were shown to have BNF ability as well as PGP properties. PGPR strains were also identified in genera *Klebsiella, Enterobacter, Citrobacter,* and *Pseudomonas* (Kennedy, Choudhury, & KecSkes, 2004). Some strains of these genera can be pathogenic to humans and one needs to exercise due caution in using them. Some of the strains are diazotrophs and are of interest for incorporation in inoculant formulations (see Section 5).

7. *Acetobacter. Gluconacetobacter (Acetobacter) diazotrophicus* is an acid-tolerant endophytic organism that grows best in a sucrose-rich medium such as sugarcane sap (James et al., 2000). Inoculation of sugarcane with *A. diazotrophicus* leads to greater growth rates and significantly increased sugarcane yield when it was applied in association with AM fungi (Muthukumarasamy, Revathi, & Lakshminarasimhan, 1999). [15]N tracer studies indicated that up to 60–80% of sugarcane plant N needs (equivalent

to over 200 kg N/ha/yr) can be met from BNF (Boddey, Polidoro, Resende, Alves, & Urquiaga, 2001). The family *Acetobacteriaceae* includes genera, *Acetobacter, Gluconabacter, Gluconacetobacter,* and *Acidomonas.*

8. *Clostridium.* Clostridia are obligatory anaerobic heterotrophs that are capable of fixing N_2 in the complete absence of oxygen (Kennedy & Tchan, 1992; Mishustin, Emtsev, & Lockmacheva, 1983). Clostridia can usually be isolated from rice soils (Elbadry, El-Bassel, & Elbanna, 1999) and their activity increased after returning straw to rice fields to raise the C-to-N ratio. Inoculation with clostridia can increase rice yields significantly under favorable conditions when carbon-rich compounds are incorporated into the soil (Kennedy & Tchan, 1992). Some strains of *Clostridium* sp. catalyze BNF in rice plants.

9. *Herbaspirillum. Herbaspirillum* is an endophyte that colonizes rice, maize, sorghum, and other cereals as well as crops such as sugarcane (James et al., 2000). *Herbaspirillum seropedicae* is an endophyte in roots, stems, and leaves of sugarcane. When used as an inoculant for sugarcane, *H. seropedicae* significantly increased growth yield and leaf N content, but could not substitute for urea N completely (Muthukumarasamy et al., 1999; Reis, Baldani, Baldani, & Dobereiner, 2000). *Herbaspirillum seropedicae* gave similar results upon inoculation of sorghum (Pereira, Cavalcante, Baldani, & Döbereiner, 1988) and maize (James et al., 2000). Inoculation of rice plants with *H. seropedicae* under field conditions increased shoot and root length and the yield increased considerably (Arangarasan, Palaniappan, & Chelliah, 1998; Baldani et al., 2000). The increase in yield of basmati rice went up and the atmospheric N_2 fixed by *H. seropedicae* accounted for 38% of the total N requirements of the plant for basmati rice and up to 60% of total N needs for super basmati rice (Mirza et al., 2000). As compared to this, N_2 fixation by *A. lipoferum* and *A. brasilense* in rice accounted for about 20% of the total N needs of the plant. When wheat was inoculated with *H. seropedicae,* it fixed N_2 and increased yields of grain and straw (Kennedy & Islam, 2001).

10. *Rhizobium.* Rhizobia, in addition to being the most important bacterial group involved in symbiotic nitrogen fixation in legumes, are also gaining recognition now as important PGPR organisms in non-legumes. Rhizobia do not form root nodules in non-legumes but confer a number of beneficial effects on the host plant as well as favorably influence some of the other microorganisms in the rhizosphere (Avis, Gravel, Autoun, & Tweddel, 2008; Babalola, 2010). Naturally occurring Rhizobia, isolated from nodules of tropical legumes, have also been shown to infect roots

of many non-legume species such as rice, wheat, and maize via cracks made by emerging lateral roots (Webster et al., 1997). In these situations, they act as endophytes, defined as rhizosphere-competent microbes that occupy inter- and intracellular niches in the plant without causing substantial harm to the host plant. These are some of the most active organisms in stimulating the growth of plants through biocontrol of the plant pathogens and by solubilizing P in the soil to a plant-usable form (Antoun & Prevost, 2006; Somers et al., 2004). In a study with maize, Chabot, Antoun, Kloepper, and Beauchamp (1996) used biolumines-cence from *R. leguminosarum* bv. *phaseoli* strain harboring lux genes to visualize in situ colonization of roots by rhizobia and to assess the effi-ciency with which these bacteria infect maize roots. These observations were consistent with findings on maize root colonization and infection by rhizobia (Gutierrez & Romero, 2001).

Rhizobium leguminosarum bv. *trifolii* was shown to colonize rice roots endo-phytically and in field trials was able to replace 25–33% of the N fertil-izer input and increased rice yields (Yanni et al., 2001). The numbers of the organism in rice tissues was too low for the observed BNF. Later, in greenhouse experiments, rice inoculated with this organism was shown to increase growth and yield of rice and showed increased uptake of P, N, and K. Based on ^{15}N tracer studies, the increased uptake of N was not due to BNF (Biswas, Ladha, & Dazzo, 2000; Biswas, Ladha, Dazzo, Yanni, & Rolfe, 2000). They also showed increased uptake of minerals from the soil. These results indicate that certain rhizobia can facilitate increased nitrogen uptake by the plant, using a mechanism other than BNF, and thus reduce dependence on chemical fertilizers and associated adverse effects on the environment.

3. PLANT GROWTH-PROMOTING MICROORGANISMS

Plant growth-promoting microorganisms (PGPM) are bacteria and fungi (such as mycorrhizal fungi) isolated from the rhizosphere, which when inoculated into the soil stimulate plant growth through one or more of sev-eral functional mechanisms. These include, but are not limited to, enhanc-ing phosphate uptake by the host plant, giving protection against fungal/bacterial plant pathogens, inducing systemic resistance in the host plant, producing siderophores, phytohormones, and other plant growth stimula-tory substances (see Fig. 3.1). It should be noted that a PGPR organism by definition must be rhizosphere-competent, i.e. colonize plant root and the adjacent soil and successfully compete with the complex native microflora present in the rhizosphere and confer beneficial effects on the plant growth and yield. It is generally accepted now that associative diazotrophs such as *Azospirillium,* that do not produce N_2-fixing nodules or other morphological modifications of the host plant, are PGPR. Symbiotic nitrogen-fixers in legume nodules or in nodules of *Alnus* spp. by *Frankia* (an actinomycete) are not included in PGPR (Antoun & Prevost, 2006). Also excluded from PGPR are legume-nodulating strains of β-proteobacteria such as *Burkhholderia caribensis* and *Cupriavidus (Ralstonia) taiwanensis.* A number of reviews and books related to PGPR have been published recently suggesting the level of interest in these organisms (Avis et al., 2008; Hayat et al., 2010; Lugtenberg & Kamilova, 2009; Nihorimbere et al. 2011; Ortiz-Castro et al., 2009; Siddiqui, 2006; Yang, Kloepper, & Ryu, 2009; Yanni et al., 2001; Zhuang, Chen, Shim, & Bai, 2007). PGPR directly promote plant growth and increase crop yields by facilitating increased uptake of nutrients such as phosphate and by a number of other mechanisms described in this section. PGPR synthesize and export phytohormones, which are called plant growth regulators (PGRs) as they play regulatory roles in plant growth and development. PGPR also stimulate plant growth and yields indirectly by protecting the plant root surface (rhizoplane) from colonization by plant pathogenic bacteria and fungi through competition for attachment sites on the rhizoplane and by the production of antimicrobial agents (Dobbelaere, Vanderleyden, & Okon, 2003; Khalid, Arshad, Shaharoona, & Mahmood, 2009; Vessey, 2003). Selected examples of PGPR are listed in Table 3.1.

3.1. Phosphate-Solubilizing Microbes (PSM)

Phosphorous (P) is a major essential macronutrient-promoting plant growth and development and it is low P concentration that most often limits plant growth. An estimated 40% of the crop yields on the world's agriculture land are limited by P availability. Phosphate fertilizers are second only to nitrogen fertilizers in terms of costs to the farmer and represent a major cost for agricultural production worldwide (see Richardson, 2007; Tilak et al., 2005). Often high concentrations of P (as phosphate) are present in inorganic and organic pools of total P in the soil as well as in microbial biomass in the soil. This total P concentration in soil is in the range of 400–1200 mg/kg. Microbial biomass accounts for 2–5% of the total P. Even though total P in soil is relatively high, concentration of P in a soluble, plant-available form is low, usually about 1 mg/kg. Approximately 70–90% of the P added to soil as phosphate fertilizer is rapidly converted into insoluble form and is not available to

Table 3.1 Selected N$_2$-fixing and plant growth-promoting rhizobacteria[#]

Microbes	N-fixation*	P-solubilization*	Biocontrol*	Crop (S)	References
Alcaligenes sp.	X	X		Indian mustard and rape	Belinov, Dodd, Safronova, Hontzeas, and Davies (2007)
Azoarcus sp.★	X			Rice	Engelhard, Hurek, and Reinhold-Hurek (2000)
Azospirillum sp.	X (A)	X	X	Rice, sugarcane	Engelhard et al. (2000), Kim et al. (2011)
Azotobacter sp.	X (F)	X	X	Rice	Rodriguez and Fraga (1999)
Bacillus sp;		X	X	Wheat, strawberry, potato, legumes	Rudrappa et al. (2008), Ryu et al. (2003)
Bradyrhizobium sp.	X (S)	X	X	Legumes: shrub and tree species	Saravanan and Natrajan (2000); Shaharoona, Arshad, and Zahir (2006)
Burkholderia sp.	X (A/S)	X	X	Rice, maize, sorghum, onion, tomato	Caballero-Mellado, Martinez-Aguilar, Tenorio, Onofre, and Estrada-de los Santos (2003); Poonguzhali et al. (2005); Sessitsch et al. (2005)
Clostridium sp	X			Sorghum, sugarcane	Bowen and Rovira (1999)
Comamonas sp.	X			Kiwi	Erturk, Ercisli, Haznedar, and Cakmakci (2010)
Corynebacterium sp.		X		Rice	Engelhard et al. (2000)
Curtobacterium sp.		X		Rice	Engelhard et al. (2000)
Enterobacter sp		X		Rice, wheat, maize	Rodriguez and Fraga (1999)
Flavobacterium sp.		X		Indian mustard	Belinov et al. (2005)
Frankia sp.	X (S)	X		*Casuarina*, *Alnus*	Wheeler and Miller (1990)
Gluconacetobacter sp.	X	X	X	Sugarcane, carrot, coffee, radish, sweet potato	Cavalcante and Döbereiner (1988); Döbereiner et al. (2003); Madhaiyan, Saravanan, Bhakiya Shilba Sandal Jovi, and Lee (2004)

Organism	N-fixer	Crop	Reference
Herbaspirillum sp.	X	Rice, wheat, maize, sugarcane	Rothballer et al. (2006)
Klebsiella sp.	X	Rice, wheat, maize, sweet potato	Caballero-Mellado et al. (2003); Rosenblueth, Martinez, Silva, and Marti'nez-Romero (2004)
Methylobacterium sp.	X	Rice, wheat, *Arabidopsis* sp.	Engelhard et al. (2000)
Paenibacillus sp.	X	Sweet potato	Reiter, Burgmann, Burg, and Sessitsch (2003)
Pseudomonas sp.	X	Indian mustard and rape; Avocado; Strawberry, Arabidopsis sp Tomato and cucumber	Belimov et al. (2007), Cazorla et al. (2006), Iavicoli, Boutet, Buchala, and Metraux (2003), Kamilova et al. (2008), Pirlak and Kose (2009)
Cupriavidus (Ralstonia) sp.**	X (S)	Rice	Chen, James, Prescott, Kierans, and Sprent (2003)
Rhizobia sp.	X (S)	Legumes	Masson-Boivin, Giraud, Perret, & Batut (2009), Miransari (2011), Rodriguez and Fraga (1999), Serraj (2004)
Rhodopseudomonas sp.	X	Rice	Engelhard et al. (2000)
Stenotrophomonas sp.	X	Coffee	Nunes and de Melo (2006)

*Azovibrio restrictus, Azaspira oryza, Azonexus fungiphilus, originally included in genus Azoarcus, have now been split and placed in 3 separate genera; (A) associative nitrogen fixers, (F) free-living nitorgen fixers, (S) symbiotic nitrogen fixers.

**Cupriavidus metallidurans (renamed from Ralstonia metallidurans that was previously known as Ralstonia eutropha and Alcaligenes eutrophus)

#Blank space indicates that the characteristic is not a feature of the genus concerned.

the plant unless solubilized by phosphate-solubilizing microbes (PSM). Furthermore, phosphate fertilizers are expensive and repeated injudicious application of phosphate fertilizer adversely affects sustainability, lowers soil fertility by affecting microbial diversity, and heavy surface runoff of P and N leads to eutrophication of aquatic systems (Sharpley et al., 2003; Tilak et al., 2005). PSM contribute to lowering of the input of phosphate fertilizer and lowering of the costs of crop production. PSM play a major role in P-cycling in the soil environment (Richardson, 2007). In four Quebec soils, the numbers of PSM ranged from 2.5 to 3.0 × 10^6 cfu/g representing 26–46% of the total microbial flora in those soils (Chabot et al., 1996). Some of the PSM strains show synergistic effect in P-solubilization while some other strains do not. When plants are P-deficient, they increase their carbon allocation to the root to increase the rhizosphere microflora and increase lateral root growth and produce long root hair to gain access to a greater volume of soil (see Morgan et al., 2005).

A majority of phosphate-solubilizing microbes in the rhizosphere of plants belong to species of genera *Bacillus, Pseudomonas,* and *Rhizobium* and these three genera contain organisms that are among the most powerful phosphate-solubilizers (Avis et al., 2008; Rodriguez & Fraga, 1999; Tilak et al., 2005). Strains of genera *Burkholderia, Agrobacterium, Micrococcus, Aerobacter, Flavobacterium,* and *Erwinia* have also been shown to convert the insoluble inorganic phosphate in the soil to a soluble plant-usable form (Table 3.1; Rodriguez & Fraga, 1999; Hayat et al., 2010; Tilak et al., 2005). Mechanisms used by PGPR to solubilize the insoluble phosphate in the soil are several. These include: 1. production of organic acids such as gluconic and oxalic acids; 2. lowering of the pH; 3. chelation of cations bound to P; and 4. release of P from phosphates and phytates by the corresponding hydrolases. Gluconic acid and 2-ketogluconic acid are the main acids of PSM that are responsible for solubilization of inorganic P. However, other organic acids such as oxalic acid, citric acid, malonic acid, succinic acid, glycolic acid, and lactic acid are also involved in other P-solubilizers (see Hayat et al., 2010; Siddiqui, 2006). Some of the more common and potent P-solubilizers such as *Bacillus, Pseudomonas,* and *Rhizobium* strains, as well as less common strains of *Enterobacter, Klebsiella, Proteus, Burkholderia,* and *Serratia* use phosphatase (phosphohydrolase) for mineralizing organic phosphates in the soil. Considerable amounts of phytate (inositol phosphate) occur in soils complexed with minerals such as iron, calcium, and magnesium and have very limited solubility (Antoun, Beauchamp, Goussard, Chabot, & Lalande, 1998; Kennedy et al., 2004). Soil bacteria, such as *Bacillus* species, produce an enzyme called phytase, which hydrolyzes phytate and releases phosphate that can be utilized by the plant (for details, see Hayat et al., 2010; Morgan et al., 2005). A phosphate-solubilizing strain of R. *leguminosarum* has been shown to colonize roots of maize and lettuce and function as a PGPR in promoting the growth of these crops (Chabot et al., 1996). Among fungi, *Aspergillus* and *Penicillium* appear to be the major genera involved in phosphate solubilization in the soil. Establishment and performance of PSM in soils is adversely affected by stress conditions such as high salt, pH, and temperature. However, a *Pseudomonas* strain isolated from rhizosphere of chickpea from alkaline soils was able to solubilize P in the presence of 10% salt, ph 12, and 45 °C (see Tilak et al., 2005).

Most soils worldwide are deficient in plant-available form of P and there has been a great deal of interest in using microbial inoculants that transform the insoluble P in soil into a plant-available form of P. Use of such inoculants would be consistent with sustainable agriculture while at the same time increasing the monetary benefits to the farmer by lowering of the quantities of agrochemical fertilizers applied to the field and potential increases in crop yields. However, results of field trials have not always been encouraging even though the results were promising in laboratory trials (See Section 5 *below for additional information on inoculants*.)

3.2. Phytohormones and Cytokinins

PGPR synthesize and export phytohormones, which are also called PGRs, and play regulatory roles in plant growth and development. It has been reported that almost 90% of the microorganisms found in the rhizosphere are capable of releasing hormones for enhanced plant growth (Diallo et al., 2011; Hayat et al., 2010; Oritz-Castro et al., 2008). There are five classes of well-known PGRs: namely auxins, gibberellins, cytokinins, ethylene, and abscissic acid. Cytokinins and auxins play an important role in shaping root architecture and regulating vascular differentiation, root apical dominance, and lateral root initiation (Aloni, Aloni, Langhans, & Ullrich, 2006).

The most common, best characterized, and most physiologically active auxin in plants is indole–3–acetic acid (IAA). Almost 80% of all rhizosphere bacteria produce IAA, which is known to stimulate both rapid (e.g. increase in cell elongation) and long-term (e.g. enhancement of cell division and differentiation) responses in plants. IAA is known to be involved in root initiation and development, root hair formation, cell enlargement, organogenesis, and gene regulation (Adesemoye, Torbert, & Kloepper, 2008, 2009; Dutta & Podile, 2010; Khalid et al., 2009). A majority of *Rhizobium* isolates from the field, produce IAA and may serve as PGPR in promoting growth of non-legume plants such as wheat. IAA-producing strains of *A. brasilense* and

Bradyrhizobium japonicum were shown to stimulate early growth promotion of seedlings of corn and soybean. Auxins are said to be the most quantitatively abundant hormones produced by *Azospirillum* (Ortiz-Castro, Martinez-Trujillo, & López-Bucio, 2008). IAA mutants of *A. brasilense* that produced little or no IAA were unable to promote the formation of lateral roots of wheat seedlings, unlike the wild type strain, indicating the importance of IAA as a plant growth regulator. It is generally agreed that production of auxins (rather than nitrogen fixation) is responsible for the observed stimulation of root development and plant growth seen in Azospirillum-inoculated plants (Hayat et al., 2010). Stimulation of plant growth by IAA-producing PGPR is affected by the amount of IAA produced by the bacterium and the response observed may vary from one plant species to another (Glick, 1995; Hayat et al., 2010). In addition to IAA, indole 3-ethanol and indole butyric acid, produced by strains of P*aenibacillus* and *Azospirillum,* contribute to PGP (Baar, Bastiaans, van de Coevering, & Roelofs, 2002).

Species of genera *Rhizobium* and *Bradyrhizobium* produce molecules such as auxins, cytokinins, abscissic acid, lumichrome, riboflavin, LCOs, and vitamins that promote plant growth (Dakora, 2003; Matiru & Dakora, 2005; Raab & Lipson, 2010). Colonization and infection of roots by rhizobia have also been reported to contribute to increase in plant growth and grain yield through phytohormone production and by inhibition of plant pathogenic microbes (Adesemoye & Kloepper, 2009; Adesemoye et al., 2008, 2009; Yang et al., 2009).

Plants and plant-associated microorganisms have been found to produce over 30 growth-promoting compounds of the cytokinin group (Arshad & Frankenberger, 1997). PGPR also produce a large range of gibberellins (GA1 to GA 110) that are diterpenoids with four isoprenoid units. Gibberellins are very similar chemically but are quite different biologically. GA1 is the most active gibberellins in plants. Gibberellins in general promote cell elongation and not growth by cell division.

Ethylene, a potent regulator of plant growth, development and senescence, is synthesized by many species of rhizobacteria. PGPR promote plant growth by keeping the level of ethylene in the plant low, thus preventing the harmful influence of high ethylene concentrations such as inhibition of root elongation as well as SNF activity in legumes (see Glick, 1995).

Cytokinins are extraordinarily active compounds produced by rhizosphere microbes. The basic nucleus called kinetin is a 6-(2-furfuryl-7-amnopurine) and a wide range of cytokinins are known (see Garcia de Salmone et al. 2010; Lugtenberg & Kamilova, 2009). Cytokinins stimulate plant growth by cell division (and not by elongation as auxins do).

Abscissic acid is widespread in the plant kingdom. It is produced by PGPR cultures of *Azospirillum* and *Rhizobium* and plays a primary role in stomatal closure. It essentially reverses the effect of growth-stimulating hormones (auxins, gibberellins, and cytokinins) in several plant tissues and inhibits seed germination. It has also been suggested that abscissic acid could be important for plant growth under water-stressed environments as in arid and semiarid climates (Hayat et al., 2010; Myresiotis, Karaoglanidis,Vryzas, & Papadopoulou-Mourkidou, 2012).

3.3. Induced Systemic Resistance

Some biocontrol agents react with plant roots and elicit induced systemic resistance (ISR), similar to innate resistance to disease in humans and animals, in the plant so that it has an enhanced defensive capacity against infection by one or more plant pathogens. This ability to elicit ISR from the plant appears to be widespread among biocontrol agents. It has been demonstrated in bean, tobacco, cucumber, and other crop plants (see Siddiqui, 2006; Babalola, 2010; Dutta & Podile, 2010; Gupta et al., 2000). There appears to be no direct correlation between the extent of colonization of the root by the biocontrol organism and its biocontrol potency since root colonization mutants of *Pseudomonas fluorescens* appear to elicit ISR. The development of ISR appears to be dependent on specific recognition between the ISR inducing biocontrol bacterium and the plant. Individual microbial cell components such as LPS and flagella, signal molecules, and some antifungal metabolites of biocontrol organisms have also been shown to elicit ISR and the list of such compounds is rather long and is still growing (Rudrappa, Czymmek, Pare, & Bais, 2008). Presence of *Rhizobium* spp. has been reported to activate the plant defense mechanism in the presence of pathogen resulting in production of phenolics, flavonoids, and phyto-alexins (Avis et al., 2008). For example, presence of *Rhizobium* spp. through elicitation of isoflavonoid phytoalexin, has been associated with resistance against disease by alfalfa and common bean (Avis et al., 2008; Dakora, 2003).

(Note: ISR should be distinguished from systemic acquired resistance (SAR), which is induced after inoculation with necrotizing pathogens and not biocontrol rhizobacterium. Furthermore, SAR requires salicylic acid as a signaling molecule in plants while ISR is dependent on jasmonic acid and ethylene signaling).

3.4. Biocontrol

In the context of this chapter, "biocontrol" is defined as a process in which one (or more) beneficial organism(s) in the rhizosphere unfavorably affect the survival or activity of a plant pathogen. The pathogenic ability of the pathogen is reduced or eliminated resulting in indirect promotion of plant growth. Plant diseases are responsible for worldwide losses of about 252 billion U.S. dollars per year (Agrios, 2005). The concept of "biocontrol" to protect crops against diseases by plant pathogens and increase food production is quite attractive as it eliminates or reduces the need for using agrochemicals, which are expensive and constitute a potential hazard for humans, animals, and the environment (see Siddiqui, 2006; Antoun et al. 1998; Morgan et al., 2005). Moreover, the inhibitory compounds produced by the biocontrol bacteria are biodegradable while agrochemicals used for inhibiting plant pathogens are designed to resist biodegradation. Also, while the productions of antimicrobial compounds by PGPR occur at or near the plant surface, many agrochemicals do not even reach the plant to be effective against the pathogen (Dobbelaere et al., 2003; Glick & Pasternak, 2003; Lugtenberg & Kamilova, 2009; Raupach and Kloepper. 1998). To be effective, the biocontrol agent should effectively compete with other microorganisms in the rhizosphere for nutrients and be able not only to colonize the roots but also occupy the right niche on the root.

PGPR as well as some arbuscular mycorrhiza (AM fungi) and *Trichoderma* species are effective biocontrol agents and have been successfully used for controlling fungal, bacterial, viral, and nematode diseases of plants for increasing crop productivity, as documented in a number of reviews (Harman, Howell, Viterbo, Chet, & Lorito, 2004; Hata, Kobae, & Banba, 2010; Nihorimbere et al., 2012). The ability of biocontrol agents in promoting greater productivity of crops is an indirect effect in that these biocontrol organisms decrease or eliminate the inhibiting effect of the pathogen resulting in a beneficial effect on plant growth. The use of PGPR for biocontrol is gaining momentum and a number of commercial biocontrol inoculants are available (see Table 3.2). Paulitz and Belanger (2001) reported that about 80 different biocontrol products were in use worldwide, but the actual number now may be a lot higher.

A comprehensive list of PGPR that suppress bacterial pathogens of plants has been presented (Siddiqui, 2006). Species of genera *Pseudomonas* and *Bacillus* appear to be the dominant organisms involved in inhibition of fungal pathogens of various crop plants. Less commonly, *Burkholderia, Azotobacter,* and a few other genera are encountered. The fungal pathogens inhibited include (but not limited to) species of genera *Fusarium, Rhizoctonia, Pythium, Macrophomina, Colleotrichum, Verticellum, Curvularia,* and *Botrytis* (Siddiqui, 2006). Substantial increases in productivity of various crops were observed in the presence of biocontrol agents as compared to untreated controls. For example, when *Burkholderia cepacia* R 55 and R 85 and *Pseudomonas putida* R 104 were inoculated into soils infected with *Rhizoctonia solani*, 62–78% of increase in dry weight of winter wheat was observed when compared to controls without the PGPR (de Freitas and Germida, 1991). PGPR exert their biocontrol effects on pathogens by one or more of the following mechanisms: 1. Production of siderophores; 2. Production of antimicrobial compounds and lytic enzymes; 3. Induction of systemic resistance; 4. Interference with quorum sensing; and 5. Other mechanisms that are not as well characterized.

1. Siderophores: Iron is an essential nutrient for a large majority of microbes and extreme iron deficiency is lethal. Iron is present in relatively low amounts in the rhizosphere and much of the iron present is in a ferric form (Fe^{+++}) that is poorly soluble and is not readily accessible to the microbe (Hafeez, Yasmin, Ariani, Zafar, & Malik, 2006). A number of PGPR in soils get around this problem by producing siderophores, small ferric iron-specific chelating compounds that bind Fe^{+++} with high affinity. Siderophores are essential for some PGPR in acquiring the iron and thrive in the rhizosphere. Many of the siderophores are hydroxamate compounds and less commonly catechol compounds. The siderophore binds to iron and the iron–siderophore complex binds to iron limitation-dependent receptors at the bacterial cell surface, which is then transported into the cell and ferrous iron (Fe^{++}) is released into the cytoplasm. Efficient scavenging of iron by biocontrol bacteria deprives the pathogens of the needed levels of iron and their growth is inhibited. In turn, this benefits the plant growth and yields. Siderophore production is quite common among PGPR used as biocontrol agents, especially in the pseudomonads and *Bacillus* species (Persello-Cartieaux, Nussaume, & Robaglia, 2003). Iron-chelating compounds were also shown to be produced by *Frankia, Streptomyces* sp., and several other organisms (see Hayat et al., 2010; Lugtenberg & Kamilova, 2009). Suppression of pathogens by PGPR was shown by nonactivity or lower activity in siderophoreless mutants as compared to wild-type strains. Siderophore-producing pseudomonads were shown to colonize roots of several crops resulting in the yield increases. Pseudomonads producing pyoverdin and pyochelin were shown to inhibit strains of *Pythium* spp. common fungal pathogen of plants (Siddiqui, 2006). The effectiveness of siderophore-producing biocontrol bacteria appear to depend on the amount of iron available, the amounts of siderophores produced, the sensitivity of the target plant pathogens, and various environmental factors.

Table 3.2 Web sources for selected commercially available microbial inoculants

Producers/trade name	Web link
Bacterial inoculants	
Accele-Grow-M	http://www.accelegrow.com/
AGRI-BUFFA®	http://www.agrichem.com.au/products/
VAULT® HP plus;	http://www.beckerunderwood.com/
INTEGRAL®; BioStacked®	http://www.ciat.cgiar.org/ourprograms/TropicalSoil/COMPRO/Documents/progress_report_1_compro_jan_june_2009.pdf
Legumefix	http://www.legumetechnology.co.uk/
Legume inoculants	http://www.seedquest.com/whitepages/africa/southafrica/id/s/soygro.htm
Vault® LVL	Becker Underwood, USA
USDA110	http://www.usda.gov/wps/portal/usda/usda-home
Twin N	http://www.mabiotec.com/main.php
Sumagrow®	http://sumagrow.org/Home_Page.html; http://thegrowpros.com/harvest-benefits
Endomycorrhizal inoculants	
AgBio-Endos	http://www.agbio-inc.com/agbioendos.Html
AM 120	http://www.ssseeds.com/other_products.Html
Bio/Organics	http://www.biconet.com/soil/BOmycorrhizae.html
BIOGROW Hydo-sol	http://www.hollandsgiants.com/soil.html
BioVAm	http://www.harbergraphics.com/Biovam/index.html
BuRize	http://www.biosci.com/brochure/BRZBro.pdf
Cerakinkong	http://www.cgc-jp.com/products/microbial/
Diehard™	http://www.horticulturalalliance.com/DIE-HARD_Endo_Drench.asp
Endorize	http://www.agron.co.il/en/Endorize.aspx
MYCOgold	http://www.alibaba.com/product/my100200874-100160217-0/Mycogold_Crop_Enhancer_Bio_Fertilizer_.html
Mycor	http://www.planthealthcare.co.uk/pdfs/mycor-flyer.pdf
MYCOSYM	http://www.mycosym.com/Documents/Flyer%20Olive%20and%20Verticilosis%20WEB.pdf
PRO-MIX 'BX'	http://www.premierhort.com/eProMix/Horticulture/TechnicalData/pdf/TD2-PRO-MIXBX-MYCORISE.pdf

Table 3.2 Web sources for selected commercially available microbial inoculants—cont'd

Producers/trade name	Web link
Rhizanova™	http://www.arthurclesen.com/resources/Rhizanova%20Overview%20Sheet.Pdf
Ectomycorrhizal spore inoculants	
BioGrow Blend®	http://www.terratech.net/product.asp?specific=jrjqime0
MycoApply®-Ecto	http://www.mycorrhizae.com/
Myke® Pro LF3	http://www.mycorrhizae.com/
Mycor Tree®	http://www.planthealthcare.com/

2. Antimicrobial compounds and lytic enzymes. Production of antimicrobial substances by one organism that kills or severely inhibits a plant pathogen is a mechanism used by a number of common biocontrol microbes (Duffy, Keel, & Defago, 2004; Mendes et al., 2011; Pugliese, Gullino, & Garibaldi, 2010). Evidence for the involvement of antibiotic inhibition of a given pathogen was primarily based on the fact that non–antibiotic producing mutants failed to inhibit the pathogen, while antibiotic over producing mutants (obtained by the genetic manipulation of the wild type strain) did inhibit the pathogen and were more effective in the disease suppression than the wild type. *Bacillus cereus* produces the antibiotics zwittermycin A and kanosamine (Emmert, Klimowicz, Thomas, & Handelsman, 2004; Lugtenberg & Kamilova, 2009). Other *Bacillus* species produce a wide spectrum of antifungal lipopeptide antibiotics that inhibit a number of fungal pathogens. These antibiotics include bacillomycin, surfactin, iturin, and plipastatin, and a few others (Adesemoye et al., 2008; Harman, 2011; Siddiqui, 2006). *Bacillus subtilis* and pseudomonads produce lipopeptide biosurfactants that inhibit plant pathogens. Hydrogen cyanide (HCN) production is another mechanism used by biocontrol rhizobacteria to protect plants against pathogens. Other major antimicrobials produced by biocontrol organisms are phenazines (including phenazine-1-carboxylic acid, phenazine-1- carboxamide, 2,4-diacetyl phloroglucinol as the major ones), 2-hexyl-5-propyl resorcinol, D- gluconic acid, ammonia, pyrrolnitrin, pyoluterin, oligomycin A, oomycin A, and xanthobaccin (See Lugtenberg & Kamilova, 2009; Siddiqui, 2006). Some antimicrobials such as phenazine and rhamnolipid act synergistically to inhibit *Pythium* spp.

Some biocontrol rhizobacteria inhibit fungal phytopathogens by producing fungal cell wall, degrading enzymes such as chitinase and β-1,3-glucanase (Myresiotis et al., 2012). Application of an actinomyces sp. producing beta glucanases resulted in inhibition of *Phytophthora fragariae* var. *rubi* (that causes raspberry rot). *Pythium ultimum* in the rhizosphere of beet root was inhibited by an extracellular protease of *Stenotrophomonas maltophila* (cited in Siddiqui, 2006). A strain of *B. cepacia* producing a β-1,3-glucanase was shown to degrade mycelia of fungal pathogens *P. ultimum, R. solani,* and *Sclerotium rolfsii.* Similarly, chitinolytic enzymes produced by some PGPR appear to inhibit *R. solani.*

3. Quorum sensing. Bacteria use quorum sensing to coordinate their pathogenic activities based on the local density of the bacterial population. Certain plant pathogenic bacteria use quorum sensing molecules such as N-acyl-L- homoserine lactones (AHLs) that sense high bacterial density, which in turn triggers production of pathogenic virulence factors followed by disease production by the pathogen. Biocontrol bacteria such as *B.thuringiensis* and Pseudomonads interfere with this quorum sensing of a phytopathogen by breaking down the AHLs and decrease or eliminate the disease-causing ability of the pathogen (Schippers, Bakker, & Bakker, 1987).

4. Competition. Certain biocontrol bacteria effectively compete with the pathogen for nutrients, oxygen, and colonization sites on the root surface (rhizoplane). Alabouvette, Lemanceau, & Steinberg (1993) suggested that competition of the biocontrol bacterium with the pathogen for nutrients (especially carbon) is responsible for lowering the pathogenic ability of certain plant pathogens. Ability of some rhizobacteria to compete with the pathogen depends on their ability to convert glucose to gluconic acid. This sequestering of glucose effectively puts the pathogen at a competitive disadvantage (Nihorimbere et al., 2011).

5. Other mechanisms. Certain biocontrol bacteria such as *P. fluorescens* colonize the hyphae of the fungal pathogen *F. oxysporum* and use the hyphae as a food source. Thus *P. fluorescens* uses predation as a mechanism for stimulation of plant growth. Furthermore, this organism inhibited spore germination and formation of new spores by *F. oxysporum* (Kamilova, Lamers, & Lugtenberg, 2008).

Volatile organic compounds (VOCs) such as alcohols, aldehydes, ketones, and hydrocarbons play a role in inhibiting certain fungal pathogens and may also play a role in communication between PGPR and their plant host.Takano-Kai et al. (2009) showed that PGPR species produce bioactive volatiles that have antifungal activity. VOCs produced by PGPR have also been demonstrated to play a role in ISR in the host plant. In this respect, VOCs appear to play a role similar to those PGPR that have biocontrol activity. Further data implicate VOCs as modulators of auxin homeostasis and in impacting pathways involved in plant morphogenesis (see Ortiz-Castro et al., 2009).

1-Aminocyclopropane-1-carboxylic acid (ACC) is a precursor of the plant hormone ethylene. Unlike ACC deaminase-negative mutants of *P. putida,* wild-type strains of this organism were able to promote the growth of canola seedlings roots, indicating that one of the mechanisms of plant growth stimulation used by this organism is the ability to produce ACC deaminase, resulting in lowering of the level of ethylene in the plant (see Glick, 1995; Madhaiyan, Poonguzhali, Ryu, & Sa, 2006).

Ability to degrade toxic soil pollutants is another PGPR functional mechanism for promoting plant growth. In these cases, selection of pollutant-degrading rhizobacteria that live on the root or close to the root are selected so that they can efficiently use the root exudate as their major nutrient source and protect the plant by breaking down the toxic pollutant. This is an important factor in designing an inoculant for polluted soils (Dutta & Podile, 2010; Lugtenberg & Kamilova, 2009). For example, a strain of *P. putida* effectively utilizes root exudate, and degrades naphthalene around the root so that plant seeds are protected from it. *Pseudomonas putida* also stimulates plant growth through one or more of its PGP characteristics.

The abundance of nitrogen fixing, phosphate solubilizing, plant disease-suppressing, and pollutant-degrading bacteria in the rhizosphere of crop plants has now been well documented (Adesemoye, Torbert, & Kloepper, 2009; Bashan et al., 2004; Dodd & Ruiz-Lozano, 2012). Isolation and intensive screening of PGPR from plant rhizosphere will pave the way for the development of effective bioformulations for enhancing the crop productivity.

3.5. Plant Growth Promotion by Fungi

Trichoderma is perhaps the most studied fungal genus from the standpoint of its ability to inhibiting plant pathogens. *Trichoderma* is a ubiquitous soil organism and occurs worldwide. *Trichoderma* spp. are effective in giving protection to many plants by direct inhibition of many of the common fungal pathogens of plants due to antibiosis and parasiticism. Furthermore, *Trichoderma* species use multiple mechanisms for indirect inhibition by eliciting ISR on the part of the plant (Reviewed in Avis et al., 2008; Harman et al., 2004). PGPR such as *Bacillus* species produce a wide spectrum of antifungal lipopeptide antibiotics such as bacillomycin, surfactin, iturin, and plipastatin, and a few others, which inhibit a number of fungal pathogens (reviewed in Siddiqui, 2006; Harman et al., 2004). A combination of *Trichoderma atroviride* and *P. putida* act synergistically and significantly increased root dry weight and yield of tomatoes (Gravel, Antoun, & Tweddell, 2007). Also, *Trichoderma harzianum* was shown to produce a number of plant growth stimulants, in addition to its documented biocontrol properties. *T. harzianum* inoculation resulted in increased cucumber yield by 80% and plant root area by 95% (Yedida, Srivastava, Kapulnik, & Chet, 2001). Moreover, a number of *Trichoderma* species have been reported to have the ability to solubilize insoluble phosphates in the soil (Harman et al., 2004).

Mycorrhizal species (see Section 4 below) that exhibit "biocontrol" properties against a number of common fungal pathogens of different crops has been comprehensively reviewed (see Fulton, 2011; Whipps, 2004). *Glomus* species, which are some of the most common arbuscular mycorrhizae that establish symbiosis with plants, appear to dominate the list of mycorrhizae that have biocontrol activity against a broad range of fungal pathogens.

4. MYCORRHIZAE

The term "Mycorrhiza" literally means root fungus (from Greek, Myco for "fungus" and rhiza for "root"). In practice, the term mycorriza is used to describe many different types of symbiotic associations between fungi and plants. In symbiotic association between a fungus and the roots of vascular plants, the fungus may colonize the roots of the host plant intracellularly ("endomycorrhiza") or extracellularly ("ectomycorrhiza"), the two types of mycorrhizal associations that are most commonly encountered in nature (Bonfante & Anca, 2009). Mycorrhizae colonize more than 90% of all terrestrial plants and are important components of soil microbiota in close proximity to the root. Mycorrhizas are found in many different environments and their ecological success reflects a high degree of diversity in the genetic and physiological abilities of these fungal endophytes. About 6000 species of mycorrhiza have been recognized in Asocmycotina, Basidiomycotina, and Glomeromycotina (Bonfante & Anca, 2009; Rillig & Mummey, 2006). Mycorrhiza form symbiotic associations with more than 200,000 plant species. Two major mycorrhizal groups, arbuscular mycorrhiza and ectomycorrhiza, are of importance in agriculture and forestry. Mycorrhizal fungi connect their plant hosts to the heterogeneously distributed nutrients required for their growth, enabling the flow of energy-rich compounds required for nutrient mobilization whilst simultaneously providing conduits for the translocation of mobilized products back to their hosts (Hata et al., 2010).

Plant-mycorrhizal interactions are bidirectional symbiotic interactions in that the plant provides carbohydrates and other metabolites (derived from photosynthesis) to mycorrhiza while the later in turn delivers P from organic polymers, release of P and other micronutrients from insoluble mineral particles or rock surfaces via weathering, influence carbon cycling, improve soil aggregation, protect against stress such as drought, serve as bio-control agents for the plant, help interactions with neighboring plants, and a range of possible interactions with other soil microbial groups (Bending et al., 2006; Bonfante & Anca, 2009; Schroeder & Janos, 2005). A number of reviews have appeared on mycorrhizal characteristics, physiological/eco-logical and functional roles, mycorrhizal phylogenomics to metabolomics, and beneficial plant–mycorrhiza–rhizobacteria interactions that contribute to increased plant productivity (Anderson & Cairney, 2007; Bonfante & Anca, 2009; Bonfante & Genre, 2008; Bending et al., 2006; Deakin and Broughton, 2009; Finlay, 2008; Fulton, 2011; Harrison, 2005; Hata et al., 2010; Khasa, Piche, & Cougham, 2009; Koltai & Kapulnik, 2010; Johansson et al., 2004; Jeffries, Gianinazzi, Perotto, Turnau, & Barea, 2003; Parniske, 2008; Peterson, 2004; Smith & Read, 2008; Shtark et al., 2011; Varma, 2008). Major contributions of mycorriza to plant growth and yield have been well documented. Mycorrhizas, not the roots, are the chief means for uptake of P and other mineral nutrients by the plants. Mycorrhiza, as exemplified by AM fungi, are continuously interactive with plant roots, a wide range of microorganism in the rhizosphere including PGPR, mycorrhizal helper bacteria (MHB), and plant pathogens. This collective environment together is referred to as mycorrhizosphere. Mycorrhiza and bacteria in the mycor-rhizo sphere often have mutual synergistic interactions (Azcon, 2009; Barea, Azcon, & Azcon-Aguilar, 2002, 2005; Bending et al., 2006; Finlay, 2008). A wide variety of gram-negative and gram-positive bacteria stimulate mycorrhizal association with plant root system and are generally referred to as Mycorrhizal Helper Bacteria (MHB) (Garbaye, 1994). In this section, we briefly summarize plant-mycorrhiza symbiosis and synergistic interactions among plant, Mycorrhiza, and bacterial flora in the mycorrhizo sphere.

4.1. Arbuscular Mycorrhizae

Endomycorrhizas include arbuscular (AMs), ericoid, and orchid mycor-rhizas. Arbuscular mycorrhizal fungi are soil fungi belonging to the phy-lum Glomeromycota (Cairney, 2000; Castillo & Pawlowska, 2010; Sturmer,

2012). AM fungal association with plants dates back to 460 million years, predating the Rhizobium-legume root nodule symbiosis by >300 million years (Cairney, 2000; Parniske, 2008; Selosse & Rousset, 2011; Shtark et al., 2011; Wilkinson, 2001). AM fungi are obligate symbionts and have limited saprobic ability. They are dependent on the plant for their carbon nutrition. AM fungi are the most commonest mycorrhizal group associated with plants and are found in association with angiosperms, gymnosperms, pteridophytes, and bryophytes (Dermatsev et al., 2010; Hoffman & Arnold, 2010; Sturmer, 2012). AM fungi are able to develop a symbiotic association with most terrestrial plants throughout the world and play an important role in providing the plant with P, S, N, and various micronutrients from the soil. AM fungi mobilize N and P from organic polymers, release mineral nutrients from insoluble particulate matter, and mediate plant responses to stress factors and resistance to plant pathogens. Mycorrhiza also participate in a number of beneficial interactions with various groups of soil microorganisms (Bianciotto & Bonfante, 2002; Krishna, 2005; Peterson, 2004; Tikhonovich & Provorov, 2007; Veresoglou, Menexes, & Rillig, 2012). Plants and their AM fungal microsymbionts interact in complex underground networks involving multiple partners (Barea, Azcón, & Azcón-Aguilar, 2005; Bending et al., 2006; Rosendahl, 2008). How these partners maintain a fair, two-way transfer of resources between themselves was not well understood. Recently, however, Kiers et al. (2011) were able to show that plants can detect, discriminate, and reward the best fungal partners with more carbohydrates, while the fungal partners, in turn, enforce cooperation by increasing the nutrient transfer only to those host plant roots that provide more carbohydrates. Such mutualism between partners is evolutionarily stable because control is bidirectional and partners offering the best rate of exchange are rewarded.

The hyphae of AM fungi penetrate the plant root cells to establish an intracellular symbiosis, irrespective of the plant host. AM fungal colonization of plant root cells is complex. AM fungal spores germinate to form hyphopodia (or appressoria) at the root surface, followed by inter- and intracellular penetration by hyphae and the formation of characteristic branched intracellular structures called "arbuscules" (Fig. 3.2, panels c, d, and e) and produce bladder-like vesicles (hyphal swellings in the root cortex that contain lipids and cytoplasm) inside cortical cells. Arbuscules, which give their name to the symbiosis, are considered the main site of exchange of C, P, water, and other nutrients between the symbiotic partners (Bonfante & Requena, 2011; Lee, Muneer, Avice, Jung, & Kim, 2012; Miransari, 2011; Parniske, 2008; Rillig & Mummey, 2006). The transfer of carbon from the

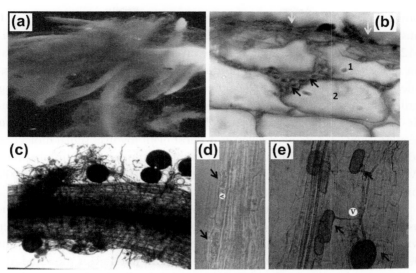

Figure 3.2 Ectomycorrhizal and endomycorrhizal colonization of *Acacia nilotica* seedlings (Saravanan, 1998; Saravanan & Natrajan, 2000). a) Morphology of a root colonized by the ectomycorrhizal fungus, *Pisolithus tinctorius*. b) Cross-section of a root colonized by *P. tinctorius*. Thick mycelial covering (mantle) on the surface of the root (yellow arrows) and intercellular hyphal network ("Hartig net", shown by red arrows). 1. epidermal cells; 2. cortical cells. c) Root surface showing germinated spores of *Glomus mosseae* (an AM fungus) d) Stained arbuscules of *G. mosseae* in the root (arrows) e) Stained vesicles of *G. mosseae* in the root (arrows). (For interpretation of the references to color in this figure legend, the reader is referred to the online version of this book.)

plant to the fungi may also occur through the intraradical hyphae. After colonization, AM fungi produce runner hyphae (extraradicular hyphae) that grow from the plant root into the soil and take up P and micronutrients, which are transferred to the plant. AM fungal hyphae, which grow from the plant root, have a high surface-to-volume ratio, making their absorptive ability greater than that of plant roots (Azcon, 2009; Bonfante & Anca, 2009). Also, these hyphae are finer than roots and can enter into pores of the soil that are inaccessible to roots. The size and the relative proportion of hyphae within the root and in the soil vary greatly between different AM species (Abdel Latef, 2011; Dumas-Gaudot et al., 2004; Lumini, Orgiazzi, Borriello, Bonfante, & Bianciotto, 2010). Furthermore, AM fungi produce another morphologically distinct type of hyphae that grow from the roots and colonize other host plants in the proximity.

The magnitude of bacteria/mycorrhizal fungus/plant symbioses depends on a number of variables such as the extent of fungal colonization, C-transfer from the plant, the amount of P in the soil, and the relative amount of P transferred to the plant (Lee et al., 2012; Sanon et al., 2012; Veresoglou et al., 2012; Zamioudis & Pieterse, 2012). The diversity of AM fungi in an ecosystem influences the diversity and productivity of plants in that environment (Adesemoye & Kloepper, 2009; Andreas & Martin, 2006; Angeles, Ouyang, Aguirre, Lammers, & Song, 2009; Bago et al., 2003; Finlay, 2008). Symbiotic N-fixation between rhizobia (including *Sinorhizobium, Bradyrhizobium, Mesorhizobium,* and *Azorhizobium*) and a host plant involves signal molecules and pathways. This symbiosis may be favorably influenced by simultaneous AM fungal colonization with the same host plant (Bonfante & Anca, 2009; Bonfante & Requena, 2011; Harrison, 2005).

Signal molecules and pathways of AM fungi and the host plant for the onset of symbiosis and the beneficial mutualism between the two have been reviewed (Akiyama & Hayashi, 2006; Bonfante & Requena, 2011; Harrison, 2005; Kiers et al., 2011). Signals are exchanged between plant roots and AM fungi that are important for plant mycorrhizal symbiosis. Recent evidence suggests that phytohormones produced by the plant including strigolactones stimulate fungal metabolism as well as hyphal branching (Bonfante & Requena, 2011; Castillo & Pawlowska, 2010; Dodd & Ruiz-Lozano, 2012). The symbiotic signals ("Myc factors") are active in small doses and stimulate the growth of the root system, and the formation of mycorrhiza association. Biotechnological processes for synthesizing large amounts of these regulator molecules have been developed and will permit the study of their effects on crops, with the objective of improving their yield, without using chemical fertilizers (Bonfante & Requena, 2011; Harrison, 2005). There are plant genes that are expressed by "Myc factors" at the onset of AM–host plant symbiosis and the fungal genes involved in structural and physiological alterations in the host plant have also been identified (Dodd & Ruiz-Lozano, 2012; Hata et al., 2010). The physiological functions of the host plant need to be modified so that the host plant can provide the fungi with the required organic carbon compounds (through root exudate) in exchange for water and nutrients provided by the fungus. Some common plant genes are expressed during the AM symbiosis as well as N-fixation by rhizobium (Akiyama & Hayashi, 2006).

4.2. Ectomycorrhizae

Ectomycorrhizal fungi (EM fungi) are phylogenetically very diverse and more than 2000 species of EM fungi worldwide have been identified, primarily from Basidiomycotina and Ascomycotina. These EM fungi form characteristic mycorrhizal associations, almost entirely with woody perennials, including Pinaceae, Betulaceae, Fagaceae, and Diperocarpaceae in tropical, subtropical, and arid environments, and are regarded as key organisms in nutrient and carbon cycles in forest ecosystems (Agerer, 2003; Becerra et al., 2005; Jakucs, Kovacs, Agerer, Romsics, & Eros-Honti, 2005). Unlike AM fungi, hyphae of EM fungi do not penetrate into the root cells but are intercellular. The hyphae penetrate into the root cortex where they form a hyphal network ("Hartig net"; see Fig. 3.2) in the intercellular space through which minerals and nutrient materials are exchanged between the fungus and the plant. The fungus forms a mantle of hyphae on the outside of the plant root that extends into the surrounding soil (Anderson & Cairney, 2007; Smith & Read, 2008). The structure of ectomycorrhizal extramatrical mycelium (extraradical mycelium) varies considerably between ectomycorrhizal species, ranging from a weft of undifferentiated mycelium around the root to a highly differentiated mycelium comprising a foraging fungal front connected to roots via rhizomorphs (Bonfante & Anca, 2009; Cairney, 2000). In angiosperm tree roots, the ectomycorrhizal hyphae penetrate the epidermal layer and spread as hyphal network (Hartig net) intercellularly (one cortical cell deep: Fig. 3.2, panels a and b), but in the case of conifers and gymnosperms the hyphal penetration reaches up to a depth of 3–4 cortical cells (Agueda, Parlade, de Miguel, & Martinez-Pena, 2006; Lupatini, Bonnassis, Steffen, Oliveira, & Antoniolli, 2008; Massicotte, Melville, & Peterson, 2005). Similar to AM fungi, ectomycorrhizae also exhibit synergistic interactions with other plant-beneficial organisms such as symbiotic N_2-fixers. For example, ectomycorrhizal symbiosis enhanced the efficiency of inoculation of two *Bradyrhizobium* strains on the growth of legumes (Andre et al., 2005). It is also of interest that similar synergies were seen when AM fungus (*Glomus mosseae*), EM fungus (*Pisolithus tinctorius*), and *Bradyrhizobium* sp. were used together to inoculate *Acacia nilotica*, enhancement of N_2 fixation, growth, and dry biomass were observed when all three organisms were present (Saravanan and Natarajan, 1996, 2000). Moreover, *Bradyrhizobium* sp. when co-inoculated with either the AM fungus or the EM fungus, gave enhancement of N_2 fixatioin as compared to the control with *Bradyrhizobium* sp. only (Fig. 3.3).

Ericoid and orchid mycorrhizas tend to be host-specific and colonize only the plants in the family Ericaceae and Orchidaceae, respectively (Bergero, Perotto, Girlanda, Vidano, & Luppi, 2000; Cairney, 2000; Wilkinson, 2001). In ericoid mycorrhizas, colonization is simple. The fungus develops inside epidermal cells, forming coils (hair-like roots enmeshed in extensive weft of hyphae) that give rise to independent infection units. Normally

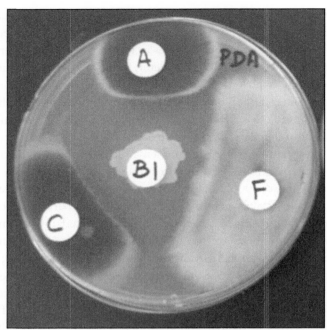

Figure 3.3 Potato dextrose agar (PDA) plate demonstrating the biocontrol property of a PGPR bacterium (B1) against three fungal pathogens A, C, and F representing strains of *Alternaria, Curvularia,* and *Fusarium* species, respectively. (For color version of this figure, the reader is referred to the online version of this book.)

no sheath is formed. Fungi that form ericoid mycorrhizal associations are all ascomycetes. These ericoid fungi play an important role in mobilizing the organic nutrients in the soil and making them available to the plant (Cairney, 2000; Smith & Read, 2008). In the ericoid type of mycorrhiza (in Ericaceae tribes Ericeae, Vaccinieae, Rhododendreae, and Calluneae and in related families), the fungus is endophytic. In the "arbutoid" type, found in members of Ericaceae tribe Arbuteae and subfamilies Pyroloideae and Monotropoideae, the association is ectendotrophic, i.e. the fungus grows within and also ensheaths the root tissue.

Orchid mycorrhizal associations involve partially or completely achlorophyllous plants (for some part of their life) and fungi of the basidiomycete group (Cairney, 2000; Smith & Read, 2008). Fungal symbionts of green orchids are highly effective saprophytes whereas those of achlorophyllous orchids are likely to form ectomycorrhizas on autotrophic plants. Coils produced by orchid mycorrhizae occur mostly in the inner layers of the root.

5. MICROBIAL INOCULANTS

To take advantage of the demonstrated beneficial effects of various soil microbial groups in increasing plant growth and yields, many different types of microbial inoculants (biofertilizers) have been in use for a long time. Biofertilizers are defined as substances containing living microbes, which when applied to seed, plant, or soil promote growth by the supply of essential nutrients such as N, P, and other mineral nutrients. In this section, the words "microbial inoculants" and "biofertilizer" are used interchangeably. There is a vast body of published research on the construction of various microbial inoculants, their applications, and the results obtained in greenhouse and field trials using these inoculants (reviewed in Arora et al., 2011; Bashan, 1998; Dodd & Ruiz-Lozano, 2012; Kennedy et al., 2004; Malusa et al., 2012; Miransari, 2011). There is growing interest in further research aimed at developing more effective, economical, and environmentally friendly microbial inoculants. The obvious reason behind this activity is the urgency to develop natural biofertilizers to meet the food needs of the increasing human population and to decrease our dependence on chemical fertilizers and pesticides.

The microbial inocula that have been in use mostly consisted of either a single organism or a combination of beneficial organisms providing one or more beneficial functions for obtaining increases in growth yields of the crops tested. Primarily, one or more of the following three major groups of microorganisms, considered beneficial to plant nutrition, were included in constructing the bioinoculants/bioformulations: 1. AM and ectomycorrhizal fungi; 2. plant growth-promoting rhizobacteria (PGPR), and 3. N_2-fixing bacteria (Feldmann, Hutter, & Schneider, 2009; Fulton, 2011; Ijdo, Cranenbrouck, & Declerck, 2011; Kennedy et al., 2004; Khan et al., 2010; Malusa et al., 2012; Vessey, 2003). Since the late 1950s, mycorrhizal fungi were utilized as biofertilizers to promote plant growth, because of their ability to increase the plant uptake of P, N, mineral nutrients, and water (Feldmann et al., 2009; Koide & Mosse, 2004; Miransari, 2011). In recent years, research efforts in the field of microbial biofertilizers have steadily increased and individual as well as various combinations of beneficial microbes conferring one or multiple beneficial effects on plant growth and yields have been described (Baldani & Baldani, 2005; Khalid et al., 2009; Krishna, 2005; Peterson, 2004). Because of the extensive literature on microbial inoculants, we have chosen to present here selected representative examples of different microbial inoculants/formulations and some preliminary work on polymicrobial inoculants.

5.1. Inocula, Carriers, and Applications

Inocula of various kinds (solid or liquid; bacterial or fungal; pure culture or mixed culture inocula, etc.) have been in use for many years. One of the first tasks to be completed in producing a microbial inoculant is to determine the desired composition of organisms to be included in the formulation. Once that is done, two other key aspects of the inoculant technology that need to be addressed are the selection of a suitable carrier that provides favorable microenvironment to keep the microbes in the formulation viable until delivery to the field.

Mycorrhizal inocula are some of the most widely used biofertilizers both in developed and developing countries. Much of the interest focused on AM fungi because of the very wide range of economically important food crops colonized by them (Jeffries et al., 2003; Khan et al., 2010; Peterson, Piche, & Plenchette, 1984). Preparation of inocula for AM Fungi is quite involved because of their obligatory requirement for plant tissue (Koide & Mosse, 2004). Commercial AMF inoculants are produced mainly by grow-ing host plants in controlled greenhouse conditions, with an inoculant containing different fungal structures (spores, hyphal fragments, and mycor-rhizal root residues) from the roots of plants used for propagation such as sorghum, maize, and onion. Substrates such as sand, soil, or other materials such as zeolite and perlite are used to mass-produce AM fungal inoculum in pots, bags, or beds (Adholeya, Tiwari, & Singh, 2005; Gianinazzi-Pearson, Séjalon-Delmas, Genre, Jeandroz, & Bonfante, 2007; Harrier & Watson, 2003; Ijdo et al., 2011). With this technique, it is possible to reach high inoc-ulum densities (Feldmann et al., 2009). The inoculum is diluted as needed prior to marketing using a carrier such as peat, clay, and fly ash.

Methods for producing monoxenic cultures of AMF were developed in the late 1980s utilizing split-plate cultures and Ri T-DNA transformed roots of carrots under controlled conditions (St-Arnaud, Hamel, Vimard, Caron, & Fortin, 1996). This method allows a higher efficiency than traditional methods with the production on the average of 15,000 spores per Petri dish 4–5 months after the beginning of the production cycle and has so far been used mainly for physiological and genetic studies in the laboratory. An improvement of this method resulted in production of about 65,000 spores in 7 months (Douds, 2002). A method for producing mycorrhizal fungi on a larger commercial scale has also been described (Adholeya et al., 2005).

Even so, the cost of producing the inoculums is somewhat high and may not be practical for field applications.

Producing ectomycorrhizal fungi is relatively easy and can be produced by fermentation using suitable media. The same is true for producing inocula of various trichoderma species. Inocula for most bacteria used in plant growth formulations is usually prepared by growing the organisms as aseptic liquid cultures, harvested, and the bacterial suspension is diluted to give a desired concentration of viable bacteria/ml (usually $\geq 10^8$/ml). When multiple organisms are used in a given formulation, each organism is grown individually and then mixed together to give a formulation with the right amount of viable bacteria per ml (Reddy and Lalithakumari, 2009a).

Carriers for preparing inocula should be designed to provide a favorable microenvironment for the PGPM to ensure their viability and adequate shelf life of the inoculant formulation (preferably two months or more at room temperature). A number of commercial inocula are sold as lyophilized products that can be reconstituted with different liquid (or solid carriers). The advantage with this is that the viability can be maintained for months or years but a large-scale preparation is more involved, careful handling is required, and the product may be relatively expensive. Maintaining viability of the microbes after soil application is just as important as maintaining viability during the storage period. Rhizosphere competence of the microbes used in preparing the inoculant is an important requirement. A desirable carrier should be easily available, stable, economical, ecofriendly, easy to apply, and have good moisture-holding capacity and pH-buffering capacity (Malusa et al., 2012). Carrier-type also determines the form of the inoculant (solid or liquid). Common solid carriers include organic materials such as peat, coal, clay, saw dust, wheat bran, peat supplemented with chitin-containing materials, and inorganic materials such as vermiculite, perlite, silicates, kaolin, and bentonite (see Adholeya et al., 2005; Kennedy et al., 2004; Malusa et al., 2012; Fig. 3.4, panel d). In the case of solid inocula, the size of the granules or beads used for immobilization of the microbe may vary from 75 μm to 250 μm. Liquid inoculants can be broth cultures, suspensions in solutions of humic acid, or suspensions in mineral or organic oils or oil-in-water suspensions. Liquid or powder-type inoculants can be used to coat the seeds, for root dipping at the time of transplantation of seedlings, or apply directly into the furrow (or seed beds) or as a foliar spray. Humic acid has been a popular carrier for a number of microbial inoculant formulations for producing a stable inoculant at ambient temperature (Reddy & Lalithakumari, 2009a, 2009b).

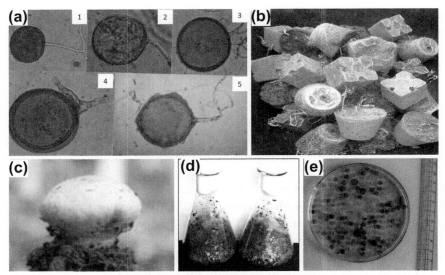

Figure 3.4 Preparation of mycorrhizal inocula for enhancing crop production (Saravanan, 1998; Saravanan & Natrajan, 2000). a) Spores of arbuscular mycorrhizae (AM) garden soil for mass inoculum production. Numbers 1 to 5 show spores of *Glomus mosseae, G. albidum, G. fasciculatum, G. macrocarpum,* and *G. dimorphicum,* respectively. b) Pot cultures of the five AM fungi mentioned in (a) above for production of spore inocula. c) Fruit body of ectomycorrhizal fungus, *Pisolithus tinctorius.* d) Mycelial inocula (45-day-old) of the ectomycorrhizal fungus *P. tinctorius.* e) *P. tinctorius* immobilized in sodium alginate beads (Mycobead). (For color version of this figure, the reader is referred to the online version of this book.)

Humic acid is a principal component of humic substances, which are the major organic constituents of soil, peat, and coal. It is primarily produced by biodegradation of dead organic matter. Humic acids consist of a mixture of weak aliphatic and aromatic organic acids that are not soluble in water under acidic conditions but are soluble under alkaline conditions. Humic acids play a vital role in soil fertility and plant nutrition. The benefits of humic acids include: stimulation of beneficial microbial activity, better seed germination, breaking up compacted soils allowing enhanced water penetration and better root growth, increased fertilizer retention, enhanced P uptake, chelation of micronutrients and increasing their bioavailability, and addition of organic matter in organically deficient soils (see R. E. Petit, web link: http://www.calciumproducts.com/articles/Dr.Pettit Humate.pdf).

Peat has been commonly used as a carrier for PGPR, particularly for rhizobial inoculants, due to its wide availability and a long history of field trials (Babalola, 2010; Dutta & Podile, 2010; Siddiqui, 2006; Somers et al., 2004).

When added to peat, PGPR maintain activity and in some cases can continue to multiply during the storage period, thus increasing their population size (Adesemoye & Kloepper, 2009). However, a major drawback of peat is the variability in its quality and composition (acidity), which affect the viability of microbes. For example, adding N-fixing and P-solubilizing bacteria increased the amount of N and P availability in the final product. Saw dust was shown to be useful as a carrier for production of inocula containing different strains of bacteria (reviewed in Arora et al., 2011; Babalola, 2010; Miransari, 2011).

Organic polymers such as alginate (D–mannuronic acid and L–glucuronic acid polymer), hydroxyethylcellulose, and carrageenan have also been suggested as inoculum carriers. Different polymers entrap or encapsulate the microorganisms, which are slowly released over a period of time. Alginate appears to be a popular carrier in this group. It is the most common polymer used for encapsulation of microbes (Fig. 3.4d). Formulations with organic polymers as carriers offer a relatively long shelf life even at ambient temperature (see Malusa et al., 2012 for details).

Application methods for the delivery of PGPM to crops in the field are relatively limited. Farmers are not keen on purchasing specialized equipment to be used for microbial–based products. Formulated inocula should be readily applied using standard farming machinery to make it appealing to farmers. Application of liquid inoculant is often more convenient for the farmer because of less time required, ease of application, and the equipment routinely used on a farm can be used. The development of inexpensive and efficient technologies for the efficient delivery of inocula, probably by modification of sprayers and sprinklers normally used for plant irrigation, could facilitate use of PGPM formulations. Soil inoculation can be done either with solid or liquid formulations. Normally, the carrier is mixed with the inoculum in the factory, but it could be mixed by the farmer prior to application, especially when liquid formulations are used. Depending on the particular inoculant formulation, the inocula can be used for seed coating, for dipping seedlings, direct application to the furrow, or as foliar application. The use of fertilizers that were produced by mixing organic matrices and insoluble phosphates with the addition of selected P-solubilizing microorganisms can also be considered for applying PGPM to crops. Such fertilizers increase the availability of nutrients (particularly of P) to plants and also improve the tolerance of the plant to pathogens (Khalid et al., 2009; Malusa et al., 2012).

Using PGPM strains that form stable and effective biofilms could be a strategy for producing commercially viable inoculant formulations (Malusa et al., 2012; Seneviratne, Zavahir, Bandara, & Weerasekara, 2008). A majority of plant-associated bacteria found on roots and in soil are found to form biofilms (Ude, Arnold, Moon, Timms-Wilson, & Spiers, 2006). Bacterial, fungal, and bacteria/fungal biofilms were suggested as possible inoculants. This is a novel and interesting idea, but to what extent this approach would be practiced remains to be seen.

5.2. Monoculture and Co-culture Inoculant Formulations

Rhizobial inoculants are extensively used around the world and the ability of rhizobia to increase plant growth and yields, resulting in a lower input of chemical fertilizers, is well established. The plant growth-promoting ability of rhizobia as inoculants varies with soil type, moisture, abundance and activity of native rhizobia, yield potential of the crop, available N in the soil, crop rotation, and other properties (Hilali, Przrost, Broughton, & Antoun, 2001). For example, an inoculant containing strains of *R. leguminosarum* bv. *trifolii,* isolated from roots of wheat cultivated in rotation with clover from loamy sand Rabat soils, gave a 24% increase ($P < 0.1$) in wheat dry biomass and grain yields, while those isolated from the silty clay Merchouch soils gave no appreciable increases in growth and yields. In trials conducted in arid areas on legumes like mung bean (*Vigna radiata*), *Bradyrhizobiuim* inoculation gave up to 10–25% yield benefits with normal rainfall (Adesemoye et al., 2009; Bashan, 1998).

Field trials with *P. fluorescens* Pf1 showed that foliar application of this organism at 7-d intervals consistently reduced the incidence of blister blight (*Exobasidium vexans*) disease in tea (*Camellia sinensis*), almost comparable in effectiveness with that of the chemical fungicide used. Also, tea yield increased significantly compared to the untreated control (Saravanakumar, Vijayakumar, Kumar, & Samiyappan, 2007). Ardakani, Heydar, Khorasani, and Arjmandi (2010) showed that biocontrol efficacy of strains of *P. fluorescens,* using bentonite or peat as a carrier, was much higher in protecting cotton seedlings against damping-off disease, as compared to controls treated with the standard carboxin-thiram fungicide. Bharathi, Vivekananthan, Harish, Ramanathan, and Samiyappan (2004) evaluated the biocontrol efficacy of 13 PGPR strains of *P. fluorescens* (Pf1) and *B. subtilis* against chilli fruit rot and die-back diseases caused by *Colletotrichum capsici,* and found them to be in increasing the seed germination and seedling vigor.

A number of studies showed that co-inoculation of two or more PGPR organism(s) gives better productivity for a range of crops. For example, legumes inoculated with *Rhizobium* and *Azospirillum* gave increase in biomass, yield, and nitrogen content. Also, early and enhanced nodulation by rhizobia co-inoculated with *Azospirillum* was attributable to an increased secretion of root flavonoid substances that are involved in the activation of the nodulation genes in *Rhizobium* (Dobbelaere et al., 2001). Stimulation of nodulation following co-inoculation with *Azospirillum* may also be due to increase in the production of lateral roots, root hair density and branching, and differentiation of a greater number of epidermal cells into root hairs, which are susceptible for infection by rhizobia. Considerable increases in yield of grain and N, P, K content were seen when wheat was coinoculated with *A. brasilense* and *Sinorhizobium meliloti* (Askary, Mostajeran, Amooaghaei, & Mostajeran, 2009; Caballero-Mellado, Carcano-Montiel, & Mascarûa-Esparza, 1992). Similar increases in N, P, and K, and various micronutrients was found in *Azospirillum-treated* maize, soybean, and rice (Caballero-Mellado et al., 1992; Naiman, Latrónico, & Garcia de Salamone, 2009; Garcia de Salamone et al., 2010; Bashan, 1998; Baldani & Baldani, 2005). More than a 100 crops and a number of environmentally important plant species were shown to be stimulated by *Azospirillum* (Bashan et al., 2004).

Trichoderma spp. and *Pseudomonas* spp. are well recognized as PGPR organisms that stimulate plant growth by multifaceted action, but primarily by their biocontrol and phosphate solubilization properties (Harman et al., 2004). The mode of action of *Trichoderma* sp. is multifaceted including antibiosis, parasitism, competition, and inducing systemic resistance (Harman et al., 2004). *Trichoderma harzianuim* increased cucumber dry mass yield by 80% when compared to the control (Yedida et al., 2001). *Pseudomonas putida* and *T. atroviride* were shown to improve both growth and fruit yields when applied to mature healthy tomato plants grown under hydroponic conditions; also, increase in the fresh weight of both the shoot and the roots of tomato seedlings was observed (Gravel et al., 2007). Multiple strains of *Trichoderma viride* and *T. harzianum* stimulate growth of lettuce. *Pseudomonas putida,* known for it is inhibition of *Fusarium* sp., was shown to increase root and shoot weight of corn (Myresiotis et al., 2012). Also, *P. putida* and *Pseudomonas cepacia* were shown to stimulate growth and yield of winter wheat (de Freitas and Germida, 1991). *Pseudomonas chlororaphis,* known as inducer of systemic resistance and also as an effective biocontrol agent, produces phenazine group of antibiotics against *Pythium aphanidermatum,* a pathogen of hot pepper seedlings. *P. chlororaphis* is also effective in eliminating soft-rot in leaves of the tobacco plant caused by *Erwinia carotovora* (see Avis et al., 2008).

Inoculation of a mixture of mycorrhiza and PGPR in general gave increased growth and yields of the crops tested, as compared to single organism inoculation (Belimov, Kojemiakor, Chuvarliyeva, 1995). AM fungi are relatively nonspecific to host plant. Co-inoculation of AM fungi with one or more of the other PGPR organisms generally gives more consistent results in enhancing growth and productivity for different crops (see Adesemoye et al., 2008; Barea et al., 2005; Dobbelaere et al., 2003; Dutta & Podile, 2010). For example, synergistic interactions between AM fungi along with a PGPR organism such as *Azospirillum, Azotobacter, Bacillus,* or *Pseudomonas* species, were found to be beneficial for enhancing plant growth and yield for a number of crops. However, the stimulatory effect of the *Azospirillum* inocula on root growth did not significantly influence the mycorrhization, regardless of the AM fungus involved, either in wheat or in maize plants, in the presence of indigenous AM fungi or when maize plants were artificially inoculated with *G. mosseae* and *Glomus macrocarpum*. Positive effects of *A. brasilense* and AM fungal colonization on rice growth and drought resistance have been reported (Ruiz-Sanchez et al., 2011). Inoculation of a mixture of mycorrhiza and PGPR in general gave increased growth and yields of the crops tested, as compared to single organism inoculation (Belimov et al., 1995). *Azospirillum-*AM fungus combination seems suitable for sustainable agriculture practices, since both types of microorganisms are not only compatible with each other but give synergistic benefits to plant productivity. Individual inoculation of *B. subtilis* and *A. brasilense* Sp245 positively affected the growth and dry weight of both shoots and roots of tomato plants, but the combination of the two rhizobacteria had no synergistic or comparable effects on plant biomass. *In vitro* tests and cellular analysis of root tips revealed growth inhibition of the primary root, which is not related to a reduced persistence in the rhizosphere of one or both bacteria (Dodd & Ruiz-Lozano, 2012). Vestberg et al. (2004) used an inoculant containing one or more strains of *G. mosseae, B. subtilis, P. fluorescens, T. harzianum,* and *Gliocladium catenalatum*. Used either singly or in dual mixtures in the presence or absence of the strawberry crown rot (caused by *Phytophthora cactorum*) and red stele (caused by *P. fragariae*), the results on decreasing disease incidence or increasing the yields have been mixed with considerable variations in treatments. These results suggest that mixing different microorganisms in the same inoculum/formulation can cause interferences and consequently give lower than expected performances. Therefore, individual organisms in mixed inoculums should be carefully chosen so that one can get synergistic increases in crop yields.

5.3. Polymicrobial Inoculant Formulations

The idea of developing a formulation containing a consortium of beneficial microbes for increasing the yields of multiple field-grown crops is not new (Antoun & Prevost, 2006; Kennedy et al., 2004), but successful formulation of this type are very few, if any. Dodd and Ruiz–Lozano (2012) suggested that combining different classes of soil organisms within one inoculant can take advantage of multiple plant growth mechanisms of such an inoculant to enhance crop productivity. However, in many studies to date, even when a mixture of PGPR and other organisms were used in designing an inoculant, the focus in many cases was on one crop rather than developing an inoculant which potentially gives positive increase in yields for multiple crops. In developing a multifunctional polymicrobial inoculant, one is expected to take into account existing crop management systems and the compatibility of the inoculant with routinely added soil organic amendments and agrochemicals. One of the main limitations so far has been the absence of a single polymicrobial inoculant formulation that would promote the growth of a range of crops including legumes, cereals, and vegetable crops.

A multistrain inoculum containing three microbial strains isolated from rice rhizosphere was used by Nguyen, Kennedy, and Roughley (2003) as a biofertilizer for rice. This multistrain biofertilizer for rice contained *Klebsiella pneuminiae, P. fluorescens* (or *P. putida), and Citrobacter freundii; P. fluorescens* was selected for its N_2-fixing ability, *Klebsiella pneumoniae* for its P–solubilizing ability, and *C. freundii,* for its ability to produce toxic substances that inhibited nearly 50% of the rice rhizosphere population (apparently to facilitate colonization by the first two organisms). Application of this biofertilizer to the rice crop allowed the reduction of urea application by 50% as compared to the uninoculated control field, which actually received twice the amount of urea fertilizer; also, 20–30% increase in rice yield was observed. The combined effect of lowering the need for urea by 50% and increasing rice yield translates into considerable cost savings for the former. Paikray and Malik (2012) reported the development of a microbial formulation containing a consortium of "*P. fluorescens, Pseudomonas striata, Paenibacillus (Bacillus) polymyxa, B. subtilis, Azospirillum, Rhizobium, Azotobacter, T. harzianum, Trichoderma viride, Saccharomyces cerevisiae* and *Lactobacillus,* and nutrients". This inoculum apparently showed a "wide variety of plant growth-promoting properties including root and shoot length elongation, early and high germination rate, high yield, decrease in soil pathogenic load and increase soil micro and macronutrient status", but details are lacking.

Polymicrobial formulations containing a diverse mixture of beneficial rhizosphere microorganisms with multiple functionalities is attractive because combining different classes of soil organisms can take advantage of multiple plant growth-promoting mechanisms and could be applied to multiple crops (Avis et al., 2008; Gravel et al., 2007; Hayat et al., 2010; Malusa et al., 2012; Vestberg et al., 2004). A key concept in constructing effective polymicrobial multifunctional formulations is the selection and use of a right combination of rhizosphere bacteria and fungi that are mutually compatible, have complementary functionalities, effectively colonize the rhizosphere of the crop(s) of interest, and bring about a synergistic promotion of growth and yield of crop(s) (Avis et al., 2008; Azcon, 2009; Barea et al., 2005; Hata et al., 2010). It is also important to select a stable and efficacious carrier that would provide a suitable microenvironment for keeping the microbes in the inoculum viable and to develop relatively simple and inexpensive delivery methods. Considering the beneficial effects of 1. PGPR; 2. AM fungi; and 3. symbiotic N_2-fixers, it is desirable to construct a polymicrobial formulation that contains several microbial strains representing each of these three functional groups and strongly promotes plant growth and yields (Arora et al., 2011; Khalid et al., 2009; Malusa et al., 2012). It is to be expected that a well-designed multifunctional formulations such as the one described would be a welcome addition to the fast-growing inoculant enterprises worldwide. Such an inoculant is also expected to be eco-friendly and suitable for organic farming and other integrated production systems, where synthetic fertilizer inputs are not allowed or restricted by law. However, construction of such complex formulations is technically demanding. To the best of our knowledge, there is no single inoculant on the market that satisfies the above mentioned criteria. Soilless and protected crops such as hydroponics can also be benefitted by such an inoculant because the predictability of the results with hydroponic systems would be higher than in open fields where a number of environmental variables may affect the outcome (Kennedy et al., 2004; Malusa et al., 2012). Ongoing efforts in the authors' lab to develop a multifunctional polymicrobial formulation of the type described above are briefly outlined below.

Several hundred bacterial strains were isolated from the root nodules of various leguminous plants as well as from soil and rhizosphere samples collected from diverse environmental sources (Reddy and Lalithakumari, 2009a, 2009b). A number of strains of *Trichoderma* species, perhaps the most-studied fungi for their inhibition of plant pathogens by antibiosis, competition, ISR, and parasitism (Harman et al., 2004), were isolated from soil samples (representing cultivated and uncultivated agricultural soils) obtained from geographically divergent regions. After screening a large number of microbial strains for the desired functionalities mentioned above, a polymicrobial formulation F2 containing phylogenetically diverse consortium of over 20 microbial strains with complementary functionalities was constructed using humate (12%, pH 7.0) as the carrier. The microbial groups in F2 were identified using 16S rDNA sequencing, and these included strains representing genera *Rhizobium, Azorhizobium, Sinorhizobium, Bacillus, Pseudomonas, Stenotrophomonas, Enterobacter,* and seven strains representing multiple species of genus *Trichoderma* (Reddy and Lalithakumari, 2009a, 2009b).

Greenhouse trials were done with various vegetables using F2 formulation as the inoculants, and the results were compared to those obtained with the controls without the added inoculant. These results are presented in Table 3.3. Substantial statistically valid increases in yields were obtained with all F2-treated plants as compared to the controls. Getting uniformly higher yields in a broad spectrum of F2-treated crops was quite encouraging. In general, F2-treated plants showed early flowering and fruiting, looked healthier, and good root nodulation was observed in legumes. Representative results are shown in Fig. 3.5. The lowest yield (37.2%) was observed with garden bean, while the highest yield (258%) was seen with okra. Although these yields are quite impressive, these experiments were done in controlled idealized conditions in a green house and the yields in a field situation would not be expected to be as high because of a number of uncontrollable variables in field experiments. Some field trials have been conducted using the F2 inoculant by BioSoil Enhancers Inc., under a license from Michigan State University. When F2 formulation, later named SumaGrow® (http://thegrowpros.com/the- science-of-sumagrow/), was used in field experiments, the yields obtained were consistent with those observed in the greenhouse experiments in that substantial increases were

Table 3.3 Green house evaluation of multifunctional polymicrobial formulation F2[#]

| | Yield (g) | | |
Crops	F2	Control	% Increase
Garden Bean (*Phaseolus vulgaris*)	48.6★	23.5	106.8
Tomato (*Lycopersicon esculentum*)	1000★	380	163.1
Wonde bush beans (*Phaseolus* sp.)	72.9★	35.6	104.8
Pea (*Pisum Sativum*)	13.9★	7.52	84.8
Pea purple hull (*Pisum arvense*)	14.75★	10.75	37.2
Okra (*Hibiscus esculentus*)	138.7★	38.7	258.4
Peanut (*Arachis hypogaea*)	21.62★	6.48	233.6
Soybeans (*Glycine max*)	11.58★	5.1	127.1

Values represent four replications for each treatment.
[#]F$_2$ formulation contained over 20 microbes with multiple functional activities such as N-fixation, phosphate solubilization, and biocontrol (see text).
★Significant, $P = 0.022$.

observed in F2-treated crops as compared to controls; however, the extent of increase in yields was less than that seen in green house experiments. For example, results of double-blind field trials showed that F2-supplemented crops gave 30%, 27%, 40%, 61%, and 75% increases in yields, respectively, for corn, bell pepper, banana pepper, yellow squash, and tomato. SumaGrow®-supplemented corn, with 50% of the normal NPK input, gave 30–40% increase in corn yields as compared to un-inoculated field with 100% NPK supplementation (http://thegrowpros.com/research/). In other words, commercial chemical fertilizer input into SumaGrow®-supplemented corn field can be reduced by 50% resulting in considerable cost savings to the farmer and safer for the environment (data from BSEI). These results are similar to the 30%–50% savings in fertilizer costs observed with some bio-inoculants by previous investigators (reviewed in Adesemoye et al., 2009; Kennedy et al., 2004; Siddiqui, 2006). To the best of our knowledge, F2 formulation is one of the very few bioinoculant products available today, that is specifically designed to contain a consortium of microbial groups with

Figure 3.5 Green house experiments testing the efficacy of polymicrobial multifunctional F2 formulation (see Text) with different plants and compared to the respective controls. a) Tomato; b) Clover; d) Switch grass. In each panel, F2-treated plants are given on the left side, while controls are given on the right side; c) Display of root nodules in garden bean inoculated with F2. (For color version of this figure, the reader is referred to the online version of this book.)

multiple complementary functions and is effective in substantially increasing the productivity of a broad spectrum of legumes, cereals, and vegetables (Reddy and Lalithakumari, 2009a, 2009b).

6. CONCLUSIONS

There is a growing worldwide awareness for the need to increase food production to feed the rapidly expanding global human population. There is also a growing consensus that the traditional approach of massive inputs of chemical fertilizers and pesticides to increase crop yields is not sustainable. Also, there is a considerable resistance in some areas of the world in using genetically engineered crops for increasing food production. Therefore, there is an obvious need for developing efficacious, environmental friendly, affordable, and safe technologies using naturally occurring rhizosphere bacteria and fungi for constructing effective microbial inoculants to increase the yields of food crops. Furthermore, successful implementation of this approach would be sustainable, nonpolluting, efficacious, and preserves soil health (Fig. 3.6).

Microbial inoculants have been in use for a long time. Symbiotic N_2-fixing rhozobia were perhaps the most widely used inoculants and are still in common use. Microbial inoculants containing one or more of the PGPR strains with or without N_2-fixing rhozobia are also in wide

Figure 3.6 Field trials demonstrating the efficacy of F2 formulation (Sumagrow®), in enhancing the growth of soybean (*Glycine max*). Left – sumagrow treated (T); right – uninoculated control (UN). (*Source: courtesy of BioSoil Enhancers Inc.*). (For color version of this figure, the reader is referred to the online version of this book.)

use for enhancing crop productivity (see Table 3.2 for a selected list of commercially available microbial inoculants). PGPR strains that are often used (singly or in mixture) include species of genera *Bacillus, Pseudomonas, Azospirillum, Azotobacter, Trichoderma,* and *Glomus;* species of other genera are used but somewhat less frequently. There is a general consensus now that these inoculants increase growth yields of various crops in the 5–30% range. There is a continuing emphasis on developing inoculants that increase yields by several different PGPR mechanisms with the expectation that such inoculants might give better yields for a broader spectrum of crops and may be more stable, affordable, and convenient to use. Preliminary laboratory and field studies with such inoculants have been encouraging.

Polymicrobial inoculants containing a large consortium of soil organisms (such as the F2 formulatioin described here) with multiple organisms representing each major PGP function such as N_2 fixation, P-solubilization, biocontrol, ISR, and production of various plant growth–promoting stimulants hold much promise for microbial enhancement of food crop productivity. There are still many unknowns and this is a productive area for intensified future research. Furthermore, in this consortium approach, a single polymicrobial formulation with multiple beneficial functions would have the ability to substantially increase productivity of a broad spectrum of food crops in diverse geographic regions. This would aid in less dependence on chemical fertilizers and pesticides, reduction of costs of farming, and protection against adverse health and environmental consequences. Also, such polymicrobial formulations consisting of microbes that naturally occur in nature, conserve soil health by increasing the number of microbial groups in the rhizosphere beneficial to crop productivity, and ensure better sustainable use of our natural resources. Effective harnessing of the power of beneficial soil organisms for alleviating human food needs appears feasible and the expanding research involving microbial physiology/ecology, biotechnology, genomics and proteomics (Tisserant et al, 2012; Wang, Ohara, Nakayashiki, Tosa, & Mayama, 2005; Weidner, Puhler, & Kuster, 2003), and applications of symbiotic N_2-fixers and PGP bacteria/fungi should yield rich dividends in the future.

ACKNOWLEDGMENTS

CAR sincerely acknowledges Dr Lalithakumari, J for her dedicated and invaluable work on the bioinoculant project in my laboratory. Acknowledgments are also due to Dr Purna Viswanathan for her help with the 16S rDNA sequencing work on the isolates.

REFERENCES

Abdel Latef, A. A. (2011). Influence of arbuscular mycorrhizal fungi and copper on growth, accumulation of osmolyte, mineral nutrition and antioxidant enzyme activity of pepper (*Capsicum annuum* L.). *Mycorrhiza, 21*, 495–503.

Adesemoye, A. O., & Kloepper, J. W. (2009). Plant-microbes interactions in enhanced fertilizer-use efficiency. *Applied Microbiology and Biotechnology, 85*, 1–12.

Adesemoye, A. O., Torbert, H. A., & Kloepper, J. W. (2008). Enhanced plant nutrient use efficiency with PGPR and AMF in an integrated nutrient management system. *Canadian Journal of Microbiology, 54*, 876–886.

Adesemoye, A. O., Torbert, H. A., & Kloepper, J. W. (2009). Plant growth-promoting rhizobacteria allow reduced application rates of chMicrobial Ecologyemical fertilizers. *Microbial Ecology, 58*, 921–929.

Adholeya, A., Tiwari, P., & Singh, R. (2005). Large-scale inoculum production of arbuscular mycorrhizal fungi on root organs and inoculation strategies. In S. Declerck, J. A. Fortin & D.-G. Strullu (Eds.), *In Vitro culture of mycorrhizas* (Vol. 4, pp. 315–338). Heidelberg: Springer Berlin.

Agerer, R. (2003). Classification of fungi in modern view. *Mycoses, 46*(Suppl. 1), 2–14.

Agrios, G. N. (2005). *Plant pathology* (5th ed.). Amsterdam: Elsevier.

Agueda, B., Parlade, J., de Miguel, A. M., & Martinez-Pena, F. (2006). Characterization and identification of field ectomycorrhizae of *Boletus edulis* and Cistus ladanifer. *Mycologia, 98*, 23–30.

Akiyama, K., & Hayashi, H. (2006). Plant and fungal signalling molecules in the arbuscular mycorrhizal symbiosis. *Tanpakushitsu Kakusan Koso, 51*, 1024–1029.

Alabouvette, C., Lemanceau, P., & Steinberg, C. (1993). Recent advances in the biological control of Fusarium wilts. *Pesticide Science, 37*, 365–373.

Aloni, R., Aloni, E., Langhans, M., & Ullrich, C. I. (2006). Role of cytokinin and auxin in shaping root architecture: regulating vascular differentiation, lateral root initiation, root apical dominance and root gravitropism. *Annals of Botany, 97*, 883–893.

Altman, A., & Hasegawa, P. M. (2012). *Plant biotechnology and agriculture: Prospects for the 21st century* (1st ed.). Amsterdam: Elsevier/Academic Press.

Anderson, I. C., & Cairney, J. W. (2007). Ectomycorrhizal fungi: exploring the mycelial frontier. *FEMS Microbiology Reviews, 31*, 388–406.

Andreas, B., & Martin, P. (2006). The most widespread symbiosis on Earth. *PLoS Biology, 4*, e239.

Andre, S., Galiana, A., Le Roux, C., Prin, Y., Neyra, M., & Duponnois, R. (2005). Ectomycorrhizal symbiosis enhanced the efficiency of inoculation with two Bradyrhizobium strains and *Acacia holosericea* growth. *Mycorrhiza, 15*, 357–364.

Angeles, J. G., Ouyang, Z., Aguirre, A. M., Lammers, P. J., & Song, M. (2009). Identification of gene interactions in fungal-plant symbiosis through discrete dynamical system modelling. *IET Systems Biology, 3*, 414–428.

Antoun, H., Beauchamp, C. J., Goussard, N., Chabot, R., & Lalande, R. (1998). Potential of *Rhizobium* and *Bradyrhizobium* species as plant growth promoting rhizobacteria on non-legumes: effect on radishes (*Raphanus sativus* L). *Plant and Soil, 204*, 57–68.

Antoun, H., & Prevost, D. (2006). Ecology of plant growth-promoting rhizobacteria. In Z. A. Siddiqui (Ed.), *PGPR: Biocontrol and biofertilization* (pp. 1–38). Dordrecht, The Netherlands: Springer.

Arangarasan, V., Palaniappan, S. P., & Chelliah, S. (1998). Inoculation effects of diazotrophs and phospho- bacteria on rice. *Indian Journal of Microbiology, 38*, 111–112.

Ardakani, S. S., Heydar, A., Khorasani, N., & Arjmandi, R. (2010). Development of new bioformulations of *Pseudomonas fluorescens* and evaluation of these products against damping-off of cotton seedlings. *Journal of Plant Pathology, 92*, 83–88.

Arora, N., Khare, E., & Maheshwari, D. (2011). Plant growth promoting rhizobacteria: constraints in bioformulation, commercialization, and future strategies. In D. K. Maheshwari (Ed.), *Plant growth and health promoting bacteria* (Vol. 18, pp. 97–116). Heidelberg: Springer Berlin.

Arshad, M., & Frankenberger, W. T., Jr. (1997). Plant growth-regulating substances in the rhizosphere: microbial production and functions. In L. S. Donald (Ed.), *Advances in agronomy* (Vol. 62, pp. 45–151). : Academic Press.

Askary, M., Mostajeran, A., Amooaghaei, R., & Mostajeran, M. (2009). Influence of the coinoculation *Azospirillum brasilense* and *Rhizobium meliloti* plus 2,4-D on grain yield and N, P, K content of *Triticum aestivum* (cv. Baccros and Mahdavi). *American-Eurasian Journal of Agriculture & Environmental Science, 5*, 296–307.

Avis, T. J., Gravel, V., Autoun, H., & Tweddel, R. J. (2008). Multifaceted beneficial effects of rhizosphere microorganisms on plant health and productivity. *Soil Biology and Biochemistry, 40*, 1733–1740.

Azcon, C. (2009). *Mycorrhizas functional processes and ecological impact.* Berlin: Springer.

Baar, J., Bastiaans, T., van de Coevering, M. A., & Roelofs, J. G. (2002). Ectomycorrhizal root development in wet Alder Carr forests in response to desiccation and eutrophication. *Mycorrhiza, 12*, 147–151.

Babalola, O. O. (2010). Beneficial bacteria of agricultural importance. *Biotechnology Letters, 32*, 1559–1570.

Bago, B., Pfeffer, P. E., Abubaker, J., Jun, J., Allen, J. W., Brouillette, J., et al. (2003). Carbon export from arbuscular mycorrhizal roots involves the translocation of carbohydrate as well as lipid. *Plant Physiology, 131*, 1496–1507.

Baldani, J. I., & Baldani, V. L. (2005). History on the biological nitrogen fixation research in graminaceous plants: special emphasis on the Brazilian experience. *Anais da Academia Brasileira de Ciencias, 77*, 549–579.

Baldani, V. L.D., Baldani, J. I., & Dobereiner, J. (2000). Inoculation of rice plants with the endophytic diazotrophs *Herbaspirillum seropedicae* and Burkholderia spp. *Biology and Fertility of Soils, 30*, 485–491.

Barea, J. M., Azcon, R., & Azcon-Aguilar, C. (2002). Mycorrhizosphere interactions to improve plant fitness and soil quality. *Antonie Van Leeuwenhoek, 81*, 343–351.

Barea, J. M., Azcón, R., & Azcón-Aguilar, C. (2005). Interactions between mycorrhizal fungi and bacteria to improve plant nutrient cycling and soil structure. In A. Varma & F. Buscot (Eds.), *Microorganisms in Soils: Roles in Genesis and Functions* (Vol. 3, pp. 195–212). Heidelberg: Springer Berlin.

Bashan, Y. (1998). Inoculants of plant growth-promoting bacteria for use in agriculture. *Biotechnology Advances, 16*, 729–770.

Bashan, Y., Holguin, G., & de-Bashan, L. E. (2004). Azospirillum-plant relationships: physiological, molecular, agricultural, and environmental advances (1997–2003). *Canadian Journal of Microbiology, 50*, 521–577.

Becerra, A., Beenken, L., Pritsch, K., Daniele, G., Schloter, M., & Agerer, R. (2005). Anatomical and molecular characterization of Lactarius aff. omphaliformis, Russula alnijorullensis and Cortinarius tucumanensis ectomycorrhizae on Alnus acuminata. *Mycologia, 97*, 1047–1057.

Belimov, A. A., Dodd, I. C., Safronova, V. I., Hontzeas, N., & Davies, W. J. (2007). *Pseudomonas brassicacearum* strain Am3 containing 1-aminocyclopropane-1-carboxylate deaminase can show both pathogenic and growth-promoting properties in its interaction with tomato. *Journal of Experimental Botany, 58*, 1485–1495.

Belimov, A. A., Hontzeas, N., Safronova, V. I., Demchinskaya, S. V., Piluzza, G., Bullitta, S., et al. (2005). Cadmium-tolerant plant growth-promoting bacteria associated with the roots of Indian mustard (*Brassica juncea* L. Czern.). *Soil Biology and Biochemistry, 37*, 241–250.

Belimov, A. A., Kojemiakor, A. P., & Chuvarliyeva, C. V., (1995). Interactions between barley and mixed cultures of nitrogen fixing and phosphate solubilizing bacteria. *Plant Soil, 173*, 29–37.

Bending, G. D., Aspray, T. J., & Whipps, J. M. (2006). Significance of microbial interactions in the mycorrhizosphere. *Advances in Applied Microbiology, 60*, 97–132.

Bergero, R., Perotto, S., Girlanda, M., Vidano, G., & Luppi, A. M. (2000). Ericoid mycorrhizal fungi are common root associates of a Mediterranean ectomycorrhizal plant (Quercus ilex). *Molecular Ecology, 9*, 1639–1649.

Bharathi, R., Vivekananthan, R., Harish, S., Ramanathan, A., & Samiyappan, R. (2004). Rhizobacteria- based bio-formulations for the management of fruit rot infection in chillies. *Crop Protection, 23*, 835–843.

Bianciotto, V., & Bonfante, P. (2002). Arbuscular mycorrhizal fungi: a specialised niche for rhizospheric and endocellular bacteria. *Antonie Van Leeuwenhoek, 81*, 365–371.

Biswas, J. C., Ladha, J. K., & Dazzo, F. B. (2000a). Rhizobia inoculation improves nutrient uptake and growth of lowland rice. *Soil Science Society of America Journal, 64*, 1644–1650.

Biswas, J. C., Ladha, J. K., Dazzo, F. B., Yanni, Y. G., & Rolfe, B. G. (2000b). Rhizobial inoculation influences seedling vigour and yield of rice. *Agronomy Journal, 92*, 880–886.

Boddey, R. M., Polidoro, J. C., Resende, A. S., Alves, B. J.R., & Urquiaga, S. (2001). Use of the ^{15}N natural abundance technique for the quantification of the contribution of N_2 fixation to sugar cane and other grasses. *Australian Journal of Plant Physiology, 28*, 889–895.

Bonfante, P., & Anca, I. A. (2009). Plants, mycorrhizal fungi, and bacteria: a network of interactions. *Annual Review of Microbiology, 63*, 363–383.

Bonfante, P., & Genre, A. (2008). Plants and arbuscular mycorrhizal fungi: an evolutionary-developmental perspective. *Trends in Plant Science, 13*, 492–498.

Bonfante, P., & Requena, N. (2011). Dating in the dark: how roots respond to fungal signals to establish arbuscular mycorrhizal symbiosis. *Current Opinion in Plant Biology, 14*, 451–457.

Bowen, G. D., & Rovira, A. D. (1999). The rhizosphere and its management to improve plant growth. *Advances in Agronomy, 66*, 1–102.

Caballero-Mellado, J., Carcano-Montiel, M. G., & Mascarûa-Esparza, M. A. (1992). Field inoculation of wheat (*Triticum aestivum*) with *Azospirillum brasilense* under temperate climate. *Symbiosis, 13*, 243–253.

Caballero-Mellado, J., Martinez-Aguilar, L., Tenorio, S., Onofre, J., Estrada-de los Santos, P. (2003). *Characterization of plant-associated N_2-fixing Burkholderia, and their potential use in agriculture. XIInt. Mol. Plant-Microbe Interact. Congress*, St. Petersburg, (pp. 80).

Cairney, J. W. (2000). Evolution of mycorrhiza systems. *Naturwissenschaften, 87*, 467–475.

Castillo, D. M., & Pawlowska, T. E. (2010). Molecular evolution in bacterial endosymbionts of fungi. *Molecular Biology and Evolution, 27*, 622–636.

Cavalcante, V. A., & Döbereiner, J. (1988). A new acid-tolerant nitrogen-fixing bacterium associated with sugarcane. *Plant and Soil, 108*, 23–31.

Cazorla, F. M., Duckett, S. B., Bergstrom, F. T., Noreen, S., Odik, R., et al. (2006). Biocontrol of avocado Dematophora root rot by the antagonistic *Pseudomonas fluorescens* PCL 1606 correlates with the production 2-hexyl-5-propyl resorcinol. *Molecular Plant–Microbe Interactions, 19*, 418–428.

Chabot, R., Antoun, H., Kloepper, J. W., & Beauchamp, C. J. (1996). Root colonization of maize and lettuce by bioluminescent *Rhizobium leguminosarum* biovar phaseoli. *Applied and Environmental Microbiology, 62*, 2767–2772.

Chen, W. M., James, E. K., Prescott, A. R., Kierans, M., & Sprent, J. I. (2003). Nodulation of Mimosa spp. by the beta-proteobacterium *Ralstonia taiwanensis*. *Molecular Plant–Microbe Interactions, 16*, 1051–1061.

Cooper, J. E. (2004). Multiple responses of rhizobia to flavonoids during legume root infection. In J. A. Callow (Ed.), *Advances in botanical research* (Vol. 41, pp. 1–62): Academic Press.

Cooper, J. E. (2007). Early interactions between legumes and rhizobia: disclosing complexity in a molecular dialogue. *Journal of Applied Microbiology, 103*, 1355–1365.

Cooper, J. E., & Scherer, H. W. (2012). Nitrogen fixation. In M. Petra (Ed.), *Marschner's mineral nutrition of higher plants* (pp. 389–408). San Diego: Academic Press.

Dakora, F. D. (2003). Defining new roles for plant and rhizobial molecules in sole and mixed plant cultures involving symbiotic legumes. *New Phytologist, 158,* 39–49.

de Freitas, J. R. d., & Germida, J. J. (1991). *Pseudomonas* cepacia and Pseudomonas putida as winter wheat inoculants for biocontrol of Rhizoctonia solani. *Canadian Journal of Microbiology, 37,* 780–784.

Deakin, W. J., & Broughton, W. J. (2009). Symbiotic use of pathogenic strategies: rhizobial protein secretion systems. *Nature Reviews Microbiology, 7,* 312–320.

Dénarié, J., & Cullimore, J. (1993). Lipo-oligosaccharide nodulation factors: a new class of signaling molecules mediating recognition and morphogenesis. *Cell, 74,* 951–954.

Dermatsev, V., Weingarten-Baror, C., Resnick, N., Gadkar, V., Wininger, S., Kolotilin, I., et al. (2010). Microarray analysis and functional tests suggest the involvement of expansins in the early stages of symbiosis of the arbuscular mycorrhizal fungus *Glomus intraradices* on tomato (Solanum lycopersicum). *Molecular Plant Pathology, 11,* 121–135.

Dessaux, Y., Hinsinger, P., & Lemanceau, P. (2010). *Plants and soil.* Berlin Heidelberg: Springer.

Diallo, S., Crepin, A., Barbey, C., Orange, N., Burini, J. F., & Latour, X. (2011). Mechanisms and recent advances in biological control mediated through the potato rhizosphere. *FEMS Microbiology Ecology, 75,* 351–364.

Dilworth, M., & Glenn, A. (1984). How does a legume nodule work? *Trends in Biochemical Science, 9,* 519–523.

Dixon, R., & Kahn, D. (2004). Genetic regulation of biological nitrogen fixation. *Nature Reviews Microbiology, 2,* 621–631.

Dobbelaere, S., Croonenborghs, A., Thys, A., Ptacek, D., Vanderleyden, J., Dutto, P., et al. (2001). Responses of agronomically important crops to inoculation with *Azospirillum*. *Australian Journal of Plant Physiology, 28,* 871–879.

Dobbelaere, S., Vanderleyden, J., & Okon, Y. (2003). Plant growth-promoting effects of diazotrophs in the rhizosphere. *Critical Reviews in Plant Sciences, 22,* 107–149.

Dodd, I. C., & Ruiz-Lozano, J. M. (2012). Microbial enhancement of crop resource use efficiency. *Current Opinion in Biotechnology, 23,* 236–242.

Douds, D. (2002). Increased spore production by *Glomus intraradices;* in the split-plate monoxenic culture system by repeated harvest, gel replacement, and resupply of glucose to the mycorrhiza. *Mycorrhiza, 12,* 163–167.

Duffy, B., Keel, C., & Defago, G. (2004). Potential role of pathogen signaling in multitrophic plant- microbe interactions involved in disease protection. *Applied and Environmental Microbiology, 70,* 1836–1842.

Dumas-Gaudot, E., Valot, B., Bestel-Corre, G., Recorbet, G., St-Arnaud, M., Fontaine, B., et al. (2004). Proteomics as a way to identify extra- radicular fungal proteins from *Glomus intraradices-* RiT-DNA carrot root mycorrhizas. *FEMS Microbiology Ecology, 48,* 401–411.

Dutta, S., & Podile, A. R. (2010). Plant growth promoting rhizobacteria (PGPR): the bugs to debug the root zone. *Critical Reviews In Microbiology, 36,* 232–244.

Elbadry, M., El-Bassel, A., & Elbanna, K. (1999). Occurrence and dynamics of phototrophic purple nonsulphur bacteria compared with other asymbiotic nitrogen fixers in rice fields of Egypt. *World Journal of Microbiology & Biotechnology, 5,* 359–362.

Emerich, D. W., & Krishnan, H. B. (Eds.), (2009). *Nitrogen fixation in crop production.* Madison, WI, USA: American Society of Agronomy.

Emmert, E., Klimowicz, A., Thomas, M., & Handelsman, J. (2004). Genetics of zwittermicin A production by Bacillus cereus. *Applied and Environmental Microbiology, 70,* 104–113.

Engelhard, M., Hurek, T., & Reinhold-Hurek, B. (2000). Preferential occurrence of diazotrophic endophytes, *Azoarcus* spp., in wild rice species and land races of *Oryza sativa* in comparison with modern races. *Environmental Microbiology, 2,* 131–141.

Erturk, Y., Ercisli, S., Haznedar, A., & Cakmakci, R. (2010). Effects of plant growth promoting rhizobacteria (PGPR) on rooting and root growth of kiwifruit (*Actinidia deliciosa*) stem cuttings. *Biological Research, 43*, 91–98.

Fasciglione, G., Casanovas, E. M., Yommi, A., Sueldo, R. J., Barassi, C. A. (2012). *Azospirillum* improves lettuce growth and transplant under saline conditions. Journal of the Science of Food and Agriculture, *92*, 2518–2523.

Feldmann, F., Hutter, I., & Schneider, C. (2009). Best production practice of arbuscular mycorrhizal inoculum. In A. Varma & A. C. Kharkwal (Eds.), *Symbiotic Fungi: Principles and practice* (Vol. 18, pp. 319–336). Berlin Heidelberg: Springer.

Fibach-Paldi, S., Burdman, S., & Okon, Y. (2012). Key physiological properties contributing to rhizosphere adaptation and plant growth promotion abilities of Azospirillum brasilense. *FEMS Microbiology Letters, 326*, 99–108.

Finlay, R. D. (2008). Ecological aspects of mycorrhizal symbiosis: with special emphasis on the functional diversity of interactions involving the extraradical mycelium. *Journal of Experimental Botany, 59*, 1115–1126.

Fulton, S. M. (2011). *Mycorrhizal fungi: soil, agriculture, and environmental implications.* Hauppauge, New York: Nova Science Publishers. (pp).

Gangwar, M., & Kaur, G. (2009). Isolation and characterization of endophytic bacteria from endo- rhizosphere of sugarcane and ryegrass. *International Journal of Microbiology, 7*(1). (on line journal).

Garbaye, J. (1994). Transley review No. 76, Helper bacteria: a new dimension to the micorrhizal symbiosis. *New Phytologist, 128*, 197–210.

Garcia de Salamone, I. E., Di Salvo, L. P., Escobar Ortega, J. S., Boa Sorte, M. P., Urquiaga, S., & Dos Santos Teixeira, K. R. (2010). Field response of rice paddy crop to inoculation with *Azospirillum*: physiology of rhizosphere bacterial communities and the genetic diversity of endophytic bacteria in different parts of the plants. *Plant & Soil, 336*, 351–362.

Gepts, P. L. (2012). *Biodiversity in agriculture: domestication, evolution, and sustainability.* Cambridge: Cambridge University Press.

Gianinazzi-Pearson, V., Séjalon-Delmas, N., Genre, A., Jeandroz, S., & Bonfante, P. (2007). Plants and arbuscular mycorrhizal fungi: cues and communication in the early steps of symbiotic interactions. In K. Jean-Claude & D. Michel (Eds.), *Advances in botanical research* (Vol. 46, pp. 181–219): Academic Press.

Glick, B. R. (1995). The enhancement of plant growth by free living bacteria. *Canadian Journal of Microbiology, 41*, 109–117.

Glick, B. R., & Pasternak, J. J. (2003). Plant growth promoting bacteria. In B. R. Glick & J. J. Pasternak (Eds.), *Molecular biotechnology principles and applications of recombinant DNA* (3rd ed.). (pp. 436–454). Washington D.C: ASM Press.

Gravel, V., Antoun, H., & Tweddell, R. J. (2007). Growth stimulation and growth yield improvement of green house tomato plants by inoculation with *Pseudomonas putida* and *Trichoderma atraviride*: possible role of indole acetic acid (IAA). *Soil Biology and Biochemistry, 39*, 1968–1977.

Gupta, A., Gopal, M., & Tilak, K. V. (2000). Mechanism of plant growth promotion by rhizobacteria. *Indian Journal of Experimental Biology, 38*, 856–862.

Gutierrez, R. A. (2012). Systems biology for enhanced plant nitrogen nutrition. *Science, 336*, 1673–1675.

Gutierrez, R. A., & Romero, E. (2001). Natural endophytic association between *Rhizobium etli* and maize (*Zea mays* L.). *Journal of Biotechnology, 91*, 117–126.

Gyaneshwar, P., Hirsch, A. M., Moulin, L., Chen, W. M., Elliott, G. N., Bontemps, C., et al. (2011). Legume- nodulating betaproteobacteria: diversity, host range, and future prospects. *Molecular Plant–Microbe Interactions, 24*, 1276–1288.

Hafeez, F. Y., Yasmin, S., Ariani, D., Zafar, Y., & Malik, K. A. (2006). Plant growth promoting bacteria as biofertilizer. *Agronomy for Sustainable Development, 26*, 143–150.

Harman, G. E. (2011). Multifunctional fungal plant symbionts: new tools to enhance plant growth and productivity. *New Phytologist, 189,* 647–649.

Harman, G. E., Howell, C. R.,Viterbo,A., Chet, I., & Lorito, M. (2004). *Trichoderma* species-opportunistic, avirulent plant symbionts. *Nature Reviews Microbiology, 2,* 43–56.

Harrier, L. A., & Watson, C. A. (2003).The role of arbuscular mycorrhizal fungi in sustainable cropping systems. *Advances in agronomy* (Vol. 79, pp. 185–225). New York:Academic Press.

Harrison, M. J. (2005). Signaling in the arbuscular mycorrhizal symbiosis. *Annual Review of Microbiology, 59,* 19–42.

Hata, S., Kobae,Y., & Banba, M. (2010). Interactions between plants and arbuscular mycorrhizal fungi. In W. J. Kwang (Ed.), *International review of cell and molecular biology* (Vol. 281, pp. 1–48). :Academic Press.

Hayat, R.,Ali, S.,Amara, U., Khalid, R., & Ahmed, I. (2010). Soil beneficial bacteria and their role in plant growth promotion: a review. *Annals of Microbiology, 60,* 579–598.

Hegazi, N. A., Fayez, M., Amin, G., Hamza, M. A., Abbas, M.,Youssef, H., et al. (1998). Diazotrophs associated with non-legumes grown in sandy soils. In K. A. Malik, M. S. Mirza & J. K. Ladha (Eds.), *Nitrogen fixation with non- legumes* (pp. 209–222). Dordrecht: Kluwer Academic Publishers.

Herschkovitz,Y., Lerner,A., Davidov,Y., Rothballer, M., Hartmann,A., Okon,Y., et al. (2005). Inoculation with the plant-growth-promoting rhizobacterium *Azospirillum brasilense* causes little disturbance in the rhizosphere and rhizoplane of maize (Zea mays). *Microbial Ecology, 50,* 277–288.

Hilali,A., Przrost, D., Broughton,W. J., & Antoun,A. (2001). Effects de l'inoculation avec des souches de *Rhizobium leguminosarum* bv. *trifolii* sur la croissance du bl'e dans deux sols du Marco. *Canadian Journal of Microbiology, 47,* 590–593.

Hoffman, M.T., & Arnold,A. E. (2010). Diverse bacteria inhabit living hyphae of phylogenetically diverse fungal endophytes. *Applied and Environmental Microbiology, 76,* 4063–4075.

Iavicoli, A., Boutet, E., Buchala, A., & Metraux, J. P. (2003). Induced systemic resistance in *Arabidopsis thaliana* in response to root inoculation with *Pseudomonas fluorescens. Molecular Plant–Microbe Interactions, 16,* 851–858.

Ijdo, M., Cranenbrouck, S., & Declerck, S. (2011). Methods for large-scale production of AM fungi: past, present, and future. *Mycorrhiza, 21,* 1–16.

Iruthayaraj, M. R. (1981). Let *Azotobacter* supply nitrogen to cotton. *Intensive Agriculture, 19,* 23.

Jakucs, E., Kovacs, G. M.,Agerer, R., Romsics, C., & Eros-Honti, Z. (2005). Morphological-anatomical characterization and molecular identification of *Tomentella stuposa* ectomycorrhizae and related anatomotypes. *Mycorrhiza, 15,* 247–258.

James, E. K., Gyaneshwar, P., Barraquio,W. L., Mathan, N., & Ladha, J. K. (2000). Endophytic diazotrophs associated with rice. In J. K. Ladha & P. M. Reddy (Eds.), *The quest for nitrogen fixation in rice* (pp. 119–140). Los Bands: International Rice Research Institute.

Jeffries, P., Gianinazzi, S., Perotto, S.,Turnau, K., & Barea, J.-M. (2003).The contribution of arbuscular mycorrhizal fungi in sustainable maintenance of plant health and soil fertility. *Biology and Fertility of Soils, 37,* 1–16.

Johansson, J. F., Paul, L. R., & Finlay, R. D. (2004). Microbial interactions in the mycorrhizosphere and their significance for sustainable agriculture. *FEMS Microbiology Ecology, 48,* 1–13.

Kamilova, F., Lamers, G., & Lugtenberg, B. (2008). Biocontrol strain *Pseudomonas fluorescens* WCS365 inhibits germination of *Fusarium oxysporum* spores in tomato root exudate as well as subsequent formation of new spores. *Environmental Microbiology, 10,* 2455–2461.

Kennedy, I. R., Choudhury,A., & KecSkes, M. L. (2004). Non-Symbiotic bacterial diazotrophs in crop- farming systems: can their potential for plant growth promotion be better exploited? *Soil Boilogy and Biochemistry, 36,* 1229–1244.

Kennedy, I. R., & Islam, N. (2001). The current and potential contribution of asymbiotic nitrogen fixation to nitrogen requirements on farms: a review. *Australian Journal of Experimental Agriculture, 41,* 447–457.

Kennedy, I. R., & Tchan, Y. (1992). Biological nitrogen fixation in non leguminous field crops: recent advances. *Plant and Soil, 141,* 93–118.

Khalid, A., Arshad, M., Shaharoona, B., & Mahmood, T. (2009). Plant growth-promoting rhizobacteria and sustainable agriculture. In M. S. Khan, A. Zaidi & J. Musarrat (Eds.), *Microbial strategies for crop improvement* (pp. 133–160). Berlin Heidelberg: Springer.

Khan, M. S., Zaidi, A., Ahmed, M., Oves, M., & Wani, P. A. (2010). Plant growth promotion by phosphate-solubilizing fungi-current perspective. *Archives of Agronomy and Soil Science, 56,* 73–98.

Khasa, D., Piche, Y., & Cougham, A. P. (2009). *Advances in mycorrhizal science and technology.* Ottawa: NRC Research Press. 197 pp.

Kiers, T., Duhamel, M., Beesetty, Y., Mensah, J. A., Franken, O., Erik Verbruggen, E., et al. (2011). Reciprocal rewards stabilize cooperation in the mycorrhizal symbiosis. *Science, 333,* 880–882.

Kim, W. I., Cho, W. K., Kim, S. N., Chu, H., Ryu, K. Y., Yun, J. C., et al. (2011). Genetic diversity of cultivable plant growth-promoting rhizobacteria in Korea. *Journal of Microbiology and Biotechnology, 21,* 777–790.

Koide, R., & Mosse, B. (2004). A history of research on arbuscular mycorrhiza. *Mycorrhiza, 14,* 145–163.

Koltai, H., & Kapulnik, Y. (2010). *Arbuscular mycorrhizas physiology and function* (2nd ed.). Dordrecht: Springer.

Krishna, K. R. (2005). *Mycorrhizas: A molecular analysis.* Enfield, N.H: Science Publishers.

Lee, G. F., & Jones, R. A. (1986). Detergent phosphate bans and eutrophication. *Environmental Science & Technology, 20,* 330–331.

Lee, B. R., Muneer, S., Avice, J. C., Jung, W. J., and Kim, T. H., (2012). Mycorrhizal colonization and P- supplement effects on N uptake and N assimilation in perennial ryegrass under well-watered and drought-stressed conditions. Mycorrhiza, *22,* 525–535.

Limpens, E., & Bisseling, T. (2003). Signaling in symbiosis. *Current Opinion in Plant Biology, 6,* 343–350.

Long, S. R. (1989). Rhizobium-legume nodulation: life together in the underground. *Cell, 56,* 203–214.

Lugtenberg, B., & Kamilova, F. (2009). Plant-growth-promoting rhizobacteria. *Annual Review of Microbiology, 63,* 541–556.

Lumini, E., Orgiazzi, A., Borriello, R., Bonfante, P., & Bianciotto, V. (2010). Disclosing arbuscular mycorrhizal fungal biodiversity in soil through a land-use gradient using a pyrosequencing approach. *Environmental Microbiology, 12,* 2165–2179.

Lupatini, M., Bonnassis, P. A. P., Steffen, R. B., Oliveira, V. L., & Antoniolli, Z. I. (2008). Mycorrhizal morphotyping and molecular characterization of *Chondrogaster angustisporus* Giachini, Castellano, Trappe & Oliveira, an ectomycorrhizal fungus from Eucalyptus. *Mycorrhiza, 18,* 437–442.

Madhaiyan, M., Poonguzhali, S., Ryu, J., & Sa, T. (2006). Regulation of ethylene levels in canola (*Brassica campestris*) by 1-aminocyclopropane-1-carboxylate deaminase-containing Methylobacterium fujisawaense. *Planta, 224,* 268–278.

Madhaiyan, M., Saravanan, V. S., Bhakiya Shilba Sandal Jovi, D., & Lee, H. S. (2004). Natural endophytic occurrence of *Gluconacetobacter diazotrophicus* in tropical and subtropical plant of Western Ghats, India. *Microbiological Research, 159,* 233–243.

Maheshwari, D. K. (2011). *Bacteria in agrobiology plant growth responses.* Berlin: Springer.

Malusa, E., Sas-Paszt, L., & Ciesielska, J. (2012). Technologies for beneficial microorganisms inocula used as biofertilizers. *Scientific World Journal,* 491206. (On line journal).

Massicotte, H. B., Melville, L. H., & Peterson, R. L. (2005). Structural features of mycorrhizal associations in two members of the Monotropoideae, *Monotropa uniflora* and Pterospora andromedea. *Mycorrhiza, 15,* 101–110.

Masson-Boivin, C., Giraud, E., Perret, X., & Batut, J. (2009). Establishing nitrogen-fixing symbiosis with legumes: how many rhizobium recipes? *Trends in Microbiology, 17,* 458–466.

Matiru, V. N., & Dakora, F. D. (2005). The rhizosphere signal molecule lumichrome alters seedling development in both legumes and cereals. *New Phytologist, 166,* 439–444.

Matthews, S. S., Sparkes, D. L., & Bullard, M. J. (2001). The response of wheat to inoculation with the diazotroph Azorhizobium caulinodans. *Aspects of Applied Biology, 63,* 35–42.

Mendes, R., Kruijt, M., de Bruijn, I., Dekkers, E., van der Voort, M., Schneider, J. H., et al. (2011). Deciphering the rhizosphere microbiome for disease-suppressive bacteria. *Science, 332,* 1097–1100.

Meyer, S. L.F., Massoud, S. I., Chitwood, D. J., & Roberts, D. P. (2000). Evaluation of *Trichoderma virens* and *Burkholderia cepacia* for antagonistic activity against root-knot nematode, Meloidogyne incognita. *Nematology, 2,* 871–879.

Miransari, M. (2011). Soil microbes and plant fertilization. *Applied Microbiology and Biotechnology, 92,* 875–885.

Mirza, M. S., Rasul, G., Mehnaz, S., Ladha, J. K., So, R. B., Ali, S., et al. (2000). Beneficial effects of inoculated nitrogen-fixing bacteria on rice. In J. K. Ladha & P. M. Reddy (Eds.), *The quest for nitrogen fixation in rice* (pp. 191–204). Los Ban~os: International Rice Res. Institute.

Mishustin, E. N., Emtsev, V. T., & Lockmacheva, R. A. (1983). Anaerobic nitrogen-fixing microorganisms of the *Clostridium* genus and their activity in soil. *Biology Bulletin, 9,* 548–558.

Morgan, J. A., Bending, G. D., & White, P. J. (2005). Biological costs and benefits to plant-microbe interactions in the rhizosphere. *Journal of Experimental Botany, 56,* 1729–1739.

Mortier, V., Holsters, M., & Goormachtig, S. (2012). Never too many? How legumes control nodule numbers. *Plant Cell and Environment, 35,* 245–258.

Muthukumarasamy, R., Revathi, G., & Lakshminarasimhan, C. (1999). Diazotrophic associations in sugar cane cultivation in South India. *Tropical Agriculture, 76,* 171–178.

Myresiotis, C. K., Karaoglanidis, G. S., Vryzas, Z., & Papadopoulou-Mourkidou, E. (2012). Evaluation of plant-growth-promoting rhizobacteria, acibenzolar-S-methyl and hymexazol for integrated control of *Fusarium* crown and root rot on tomato. *Pest Management Science, 68,* 404–411.

Naiman, A. D., Latrónico, A. E., & Garcia de Salamone, I. E. (2009). Inoculation of Wheat with *Azospirillum brasilense* and *Pseudomonas fluorescens*: impact on the production and rhizospheric microflora. *European Journal of Soil Biology, 45,* 44–51.

Nakata, P. A. (2002). The generation of a transposon-mutagenized *Burkholderia glumae* library to isolate novel mutants. *Plant Science, 162,* 267–271.

Nguyen, T. H., Kennedy, I. R., & Roughley, R. J. (2003). The response of field grown rice to inoculation with a multi-strain biofertilizer in the Hanoi district, Vietnam. In I. R. Kennedy & A. T.M.A. Choudhury (Eds.), *Biofertilizers in action* (pp. 37–44). Canberra: Rural Industries Research and Development Corporation.

Nihorimbere, V., Cawoy, H., Seyer, A., Brunelle, A., Thonart, P., & Ongena, M. (2012). Impact of rhizosphere factors on cyclic lipopeptide signature from the plant beneficial strain *Bacillus amyloliquefaciens* S499. *FEMS Microbiology Ecology, 79,* 176–191.

Nihorimbere, V., Ongena, M., Smargiassi, M., & Thonart, P., (2011). Beneficial effect of the rhizosphere microbial community for plant growth and health. *Biotechnology Agronomy Society and Environment, 15,* 327–337.

Nihorimbere, V., Ongena, M., Cawoy, H., Henry, G., Brostaux, Y., Kakana, P., et al. (2009). Bacillus-based biocontrol of fusarium disease on tomato cultures in Burundi. *Communications in Agricultural and Applied Biological Sciences, 74*, 645–649.

Nunes, F. V., & de Melo, I. S. (2006). Isolation and characterization of endophytic bacteria of coffee plants and their potential in caffeine degradation. In A. Kungolos, C. A. Brebbia, C. P. Samaras & V. Popov (Eds.), *Environmental toxicology* (Vol. 10, pp. 293–297). Southampton: Computational Mechanics Publications Ltd.

Ortiz-Castro, R., Contreras-Cornejo, H. A., Macias-Rodriguez, L., & López-Bucio, J. (2009). The role of microbial signals in plant growth and development. *Plant Signaling & Behavior, 4*, 701–712.

Ortiz-Castro, R., Martinez-Trujillo, M., & López-Bucio, J. (2008). N-acyl-L-homoserine lactones: a class of bacterial quorum-sensing signals alter post-embryonic root development in Arabidopsis thaliana. *Plant Cell and Environment, 31*, 1497–1509.

Paikray S. P. and Malik, V. S. (2012). Novel formulation of microbial consortium based bioinoculant for wide spread use in agriculture practices. Patent application, web link: http://www.faqs.org/patents/app/20120015806).

Pandey, A., & Kumar, S. (1989). Potential of *Azotobacter* and *Azospirillum* as biofertilizers for upland agriculture: a review. *Journal of Scientific and Industrial Research, 48*, 134–144.

Parniske, M. (2008). Arbuscular mycorrhiza: the mother of plant root endosymbioses. *Nature Reviews Microbiology, 6*, 763–775.

Paulitz, T. C., & Belanger, R. R. (2001). Biological control in greenhouse systems. *Annual Review of Phytopathology, 39*, 103–133.

Pawlowska, T. E., & Taylor, J. W. (2004). Organization of genetic variation in individuals of arbuscular mycorrhizal fungi. *Nature, 427*, 733–737.

Pereira, J. A. R., Cavalcante, V. A., Baldani, J. I., & Döbereiner, J. (1988). Field inoculation of sorghum and rice with *Azospirillum* spp. and Herbaspirillum seropedicae. *Plant and Soil, 110*, 269–274.

Persello-Cartieaux, F., Nussaume, L., & Robaglia, C. (2003). Tales from the underground: molecular plant-rhizobacterial interactions. *Plant Cell and Environment, 26*, 189–199.

Peterson, R. L. (2004). *Mycorrhizas: Anatomy and cell biology.* Wallingford, UK: CABI Pub.

Peterson, R. L., Piche, Y., & Plenchette, C. (1984). Mycorrhizae and their potential use in the agricultural and forestry industries. *Biotechnology Advances, 2*, 101–120.

Pirlak, M., & Kose, M. (2009). Effects of plant growth promoting rhizobacteria on yield and some fruit properties of strawberry. *Journal of Plant Nutrition, 32*, 1173–1184.

Poonguzhali, S., Madhaiyan, M., Thangaraju, M., Ryu, J. H., Chung, K. Y., & Sa, T. M. (2005). Effect of co-cultures, containing N fixer and P-solubilizer, on the growth and yield of pearl millet (*Pennisetum glaucum* (L.) R. Br.) and Blackgram (*Vigna mungo* L.). *Journal of Microbiology and Biotechnology, 15*, 903–908.

Promé, J. -C. (1996). Signaling events elicited in plants by defined oligosaccharide structures. *Current Opinion in Structural Biology, 6*, 671–678.

Pugliese, M., Gullino, M. L., & Garibaldi, A. (2010). Efficacy of microorganisms selected from compost to control soil-borne pathogens. *Communications in Agricultural and Applied Biological Sciences, 75*, 665–669.

Raab, T. K., & Lipson, D. A. (2010). The rhizosphere: a synchrotron-based view of nutrient flow in the root zone. In S. Balwant & G. Markus (Eds.), *Developments in soil science* (Vol. 34, pp. 171–198). : Elsevier.

Raupach, G. S., & Kloepper, J. W. (1998). Mixtures of plant growth-promoting rhizobacteria enhance biological control of multiple cucumber pathogens. *Phytopathology, 88*, 1158–1164.

Reddy, C. A., Lalithakumari, J. (2009a). Polymicrobial formulations for enhanced productivity of a broad spectrum of crops. Lead Papers, IVth World Congress on Conservation Agricuture, (pp. 94–101). (http ://www. fao.org/ag/ca/doc/wwcca-leadpap ers.pdf).

Reddy, C. A., Lalithakumari, J. (2009b). Polymicrobial formulations for enhancing crop productivity. U. S. Patent US2009/0308121 AI (Pending).

Reinhold-Hurek, B., Hurek, T., Gillis, M., Hoste, B., Vancanneyt, M., Kersters, K., et al. (1993). *Azoarcus* gen Nov., a nitrogen fixing Proteobacteria associated with the roots of Kallar grass (*Leptochloa fusca*) (L. Kunth.), and description of two species *Azoarcus indigens* sp. nov. and *Azoarcus communis* sp. nov. *International Journal of Systematic Bacteriology, 43*, 574–588.

Reis, V. M., Baldani, J. I., Baldani, V. L.D., & Dobereiner, J. (2000). Biological dinitrogen fixation in gramineae and palm trees. *Critical Reviews in Plant Sciences, 19*, 227–247.

Reiter, B., Burgmann, H., Burg, K., & Sessitsch, A. (2003). Endophytic *nifH* gene diversity in African sweet potato. *Canadian Journal of Microbiology, 49*, 549–555.

Richardson, A. E. (2007). Making microorganisms mobilize soil phosphorus. In E. Velazquez & C. Rodriguez-Barrueco (Eds.), *First international meeting on microbial phosphate solubilization* (Vol. 102, pp. 85–90). Netherlands: Springer.

Rillig, M. C., & Mummey, D. L. (2006). Mycorrhizas and soil structure. *New Phytologist, 171*, 41–53.

Robertson, G. P., & Vitousek, P. M. (2009). Nitrogen in agriculture: balancing the cost of an essential resource. *Annual Review of Microbiology, 63*, 541–556.

Rodriguez, H., & Fraga, R. (1999). Phosphate solubilizing bacteria and their role in plant growth promotion. *Biotechnology Advances, 17*, 319–339.

Rosenblueth, M., Martinez, L., Silva, J., & Martínez-Romero, E. (2004). *Klebsiella variicola*, a novel species with clinical and plant associated isolates. *Systematic and Applied Microbiology, 27*, 27–35.

Rosendahl, S. (2008). Communities, populations and individuals of arbuscular mycorrhizal fungi. *New Phytologist, 178*, 253–266.

Rothballer, M., Schmid, M., Klein, I., Gattinger, A., Grundmann, S., & Hartmann, A. (2006). *Herbaspirillum hiltneri* sp. nov., isolated from surface-sterilized wheat roots. *International Journal of Systematic and Evolutionary Microbiology, 56*, 1341–1348.

Rudrappa, T., Czymmek, K. J., Pare, P. W., & Bais, H. P. (2008). Root-secreted malic acid recruits beneficial soil bacteria. *Plant Physiology, 148*, 1547–1556.

Ruiz-Sanchez, M., Armada, E., Munoz, Y., Garcia de Salamone, I. E., Aroca, R., Ruiz-Lozano, J. M., et al. (2011). Azospirillum and arbuscular mycorrhizal colonization enhanced rice growth and physiological traits under well-watered and drought conditions. *Journal of Plant Physiology, 168*, 1031–1037.

Ryu, C. M., Farag, M. A., Hu, C. H., Reddy, M. S., Wei, H. X., Paré, P. W., et al. (2003). Bacterial volatiles promote growth in *Arabidopsis*. *Proceedings of National Academy of Sciences of the United States of America, 100*, 4927–4932.

Sabir, A., Yazici, M. A., Kara, Z., & Sahin, F. (2012). Growth and mineral acquisition response of grapevine rootstocks (*Vitis* spp.) to inoculation with different strains of plant growth-promoting rhizobacteria (PGPR). *Journal of the Science of Food and Agriculture, 92*, 2148–2153.

Saleh, S. A., Mekhemar, G. A. A., El-Soud, A. A. A., Ragab, A. A., & Mikhaeel, F. T. (2001). Survival of *Azorhizobium* and *Azospirillum* in different carrier materials: inoculation of wheat and Sesbania rostrata. *Bulletin of Faculty of Agriculture, Cairo University, 52*, 319–338.

Sanon, A., Beguiristain, T., Cebron, A., Berthelin, J., Sylla, S. N., & Duponnois, R. (2012). Differences in nutrient availability and mycorrhizal infectivity in soils invaded by an exotic plant negatively influence the development of indigenous *Acacia* species. *Journal of Environmental Management, 95*(Suppl.), S275–S279.

Saravanakumar, D., Vijayakumar, C., Kumar, N., & Samiyappan, R. (2007). PGPR-induced defense responses in the tea plant against blister blight disease. *Crop Protection, 26*, 556–565.

Saravanan, R.S. (1998). Studies on the effect of mycorrhizal fungi on the growth of *Acacia nilotica* seedlings in the nursery. Thesis, Doctor of philosophy, University of Madras, Madras, India.

Saravanan, R. S., & Natarajan, K. (1996). Effect of *Pisolithus tinctorius* on the nodulation and nitrogen fixing potential of *Acacia nilotica* seedlings. *Kavaka, 24*, 41–49.

Saravanan, R. S., & Natrajan, K. (2000). Effect of ecto- and endomycorrhizal fungi along with *Bradyrhizobium* sp. on the growth and nitrogen fixation in *Acacia nilotica* seedlings in the nursery. *Journal of Tropical Forest Science, 12*, 348–356.

Schippers, B., Bakker, A. W., & Bakker, P. A.H.M. (1987). Interactions of deleterious and beneficial microorganisms and the effect of cropping practices. *Annual Review of Phytopathology, 25*, 339–358.

Schroeder, M. S., & Janos, D. P. (2005). Plant growth, phosphorus nutrition, and root morphological responses to arbuscular mycorrhizas, phosphorus fertilization, and intraspecific density. *Mycorrhiza, 15*, 203–216.

Selosse, M. A., & Rousset, F. (2011). Evolution. The plant-fungal marketplace. *Science, 333*, 828–829.

Seneviratne, G., Zavahir, J., Bandara, W., & Weerasekara, M. (2008). Fungal-bacterial biofilms: their development for novel biotechnological applications. *World Journal of Microbiology & Biotechnology, 24*, 739–743.

Serraj, R. (2004). *Symbiotic nitrogen fixation: prospects for enhanced application in tropical agriculture.* Enfield, (NH): Science Publishers. XIII, 367 p.

Sessitsch, A., Coenye, T., Sturz, A.V., Vandamme, P., Ait Barka, E., Salles, J. F., et al. (2005). *Burkholderia phytofirmans* sp. nov., a novel plant-associated bacterium with plant-beneficial properties. *International Journal of Systematic and Evolutionary Microbiology, 55*, 1187–1192.

Shaharoona, B., Arshad, M., & Zahir, Z. A. (2006). Effect of plant growth promoting rhizobacteria containing ACC-deaminase on maize (*Zea mays* L.) growth under axenic conditions and on nodulation in mung bean (*Vigna radiata* L.). *Letters in Applied Microbiology, 42*, 155–159.

Sharma, H. C. (2009). *Biotechnological approaches for pest management and ecological sustainability.* Boca Raton, FL: CRC Press.

Sharpley, A. N., Daniel, T., Sims, T., Lemunyon, J., Stevens, R., & Parry, R. (2003). *Agricultural phosphorus and eutrophication* (ed.). : U.S. Department of Agriculture, Agricultural Research Service, ARS. 149.

Shtark, O., Provorov, N., Mikic, A., Alexey Borisov, A., Branko Cupina, B., & Tikhonovich, I. (2011). Legume root symbioses: natural history and prospects for improvement. *Soil Microbiology, 48*, 291–304.

Siddiqui, Z. A. (2006). *PGPR biocontrol and biofertilization.* Dordrecht: Springer.

Smith, S. E., & Read, D. J. (2008). *Mycorrhizal symbiosis* (3rd ed.). New York: Elsevier. 787 pp.

Soliman, S., Seeda, M. A., Aly, S. S.M., & Gadalla, A. M. (1995). Nitrogen fixation by wheat plants as affected by nitrogen fertilizer levels and nonsymbiotic bacteria. *Egyptian Journal of Soil Science, 35*, 401–413.

Somers, E., Vanderleyden, J., & Srinivasan, M. (2004). Rhizosphere bacterial signaling: a love parade beneath our feet. *Critical Reviews In Microbiology, 30*, 205–240.

St-Arnaud, M., Hamel, C., Vimard, B., Caron, M., & Fortin, J. A. (1996). Enhanced hyphal growth and spore production of the arbuscular mycorrhizal fungus *Glomus intraradices* in an in vitro system in the absence of host roots. *Mycological Research, 100*, 328–332.

Sturmer, S. L. (2012). A history of the taxonomy and systematics of arbuscular mycorrhizal fungi belonging to the phylum Glomeromycota. *Mycorrhiza, 22*, 247–258.

Takano-Kai, N., Jiang, H., Kubo, T., Sweeney, M., Matsumoto, T., Kanamori, H., et al. (2009). Evolutionary history of *GS3*, a gene conferring grain length in rice. *Genetics, 182*, 1323–1334.

Tikhonovich, I.A., & Provorov, N.A. (2007). Beneficial plant–microbe interactions. Chapter 14, In T. D.Yu, V. G. Dzhavakhiya, V. G.D.T. Korpela, Yu. T. Dyakov & T. Korpela (Eds.), *Comprehensive and molecular phytopathology* (pp. 365–420). Amsterdam: Elsevier.

Tilak, K. V. B. R., Ranganayaki, N., Pal, K. K., De, R., Saxena, A. K., Nautiyal, C. S., et al. (2005). Diversity of plant growth and soil health supporting bacteria. *Current Science, 89,* 136–150.

Tisserant, E., Kohler, A., Dozolme-Seddas, P., Balestrini, R., et al. (2012). The transcriptome of the arbuscular mycorrhizal fungus *Glomus intraradices* (DAOM 197198) reveals functional tradeoffs in an obligate symbiont. *New Phytologist, 193,* 755–769.

Toro, M., Azcon, R., & Barea, J. (1997). Improvement of arbuscular mycorrhiza development by inoculation of soil with phosphate-solubilizing rhizobacteria to improve rock phosphate bioavailability (sup32P) and Nutrient Cycling. *Applied and Environmental Microbiology, 63,* 4408–4412.

Ude, S., Arnold, D. L., Moon, C. D., Timms-Wilson, T., & Spiers, A. J. (2006). Biofilm formation and cellulose expression among diverse environmental *Pseudomonas* isolates. *Environmental Microbiology, 8,* 1997–2011.

Vandamme, P., Goris, J., Chen, W. -M., Vos de, P., & Willems, A. (2002). *Burkholderia tuberum* sp. nov. and *Burkholderia phymatum* sp. nov., nodulate the roots of tropical legumes. *Systematic Applied Microbiology, 25,* 507–512.

Varma, A. (2008). *Mycorrhiza: State of the art, genetics and molecular Biology, eco-function, biotechnology, eco-physiology, structure and systematics* (3rd ed.). Berlin: Springer.

Varma, A., & Kharkwal, A. C. (2009). *Symbiotic Fungi: Principles and practice.* Heidelberg: Springer.

Veresoglou, S. D., Menexes, G., & Rillig, M. C. (2012). Do arbuscular mycorrhizal fungi affect the allometric partition of host plant biomass to shoots and roots? A meta-analysis of studies from 1990 to 2010. *Mycorrhiza, 22,* 227–235.

Vessey, J. K. (2003). Plant growth promoting rhizobacteria as biofertilizers. *Plant and Soil, 255,* 571–586.

Vestberg, M., Kukkonen, S., Saari, K., Parikka, P., Huttunen, J., Tainio, L., et al. (2004). Microbial inoculation for improving the growth and health of micropropagated strawberry. *Applied Soil Ecology, 27,* 243–258.

Vincent, C., Goettel, M. S., & Lazarovits, G. (2007). *Biological control: A global perspective.* Wallingford, UK: CABI.

Wang, Y., Ohara, Y., Nakayashiki, H., Tosa, Y., & Mayama, S. (2005). Microarray analysis of the gene expression profile induced by the endophytic plant growth-promoting rhizobacteria, *Pseudomonas fluorescens* FPT9601-T5 in Arabidopsis. *Molecular Plant–Microbe Interactions, 18,* 385–396.

Wang, D., Yang, S., Tang, F., & Zhu, H. (2012). Symbiosis specificity in the legume: rhizobial mutualism. *Cell Microbiology, 14,* 334–342.

Webster, G., Gough, C., Vasse, J., Batchelor, C. A., O'Callaghan, K. J., Kothari, S. L., et al. (1997). Interactions of rhizobia with rice and wheat. *Plant and Soil, 194,* 115–122.

Weidner, S., Puhler, A., & Kuster, H. (2003). Genomics insights into symbiotic nitrogen fixation. *Current Opinion in Biotechnology, 14,* 200–205.

Wheeler, C. T., & Miller, I. M. (1990). Current and potential uses of actinorhizal plants in Europe. In C. R. Schwintzer & J. D. Tjepkema (Eds.), *The biology of frankia and Actinorhizal plants* (pp. 365–389). San Diego: Academic Press.

Whipps, J. M. (2004). Prospects and limitations for mycorrhizas in biocontrol of root pathogens. *Canadian Journal of Botany, 82,* 1198–1227.

Wilkinson, D. M. (2001). Mycorrhizal evolution. *Trends in Ecology & Evolution, 16,* 64–65.

Yang, J., Kloepper, J. W., & Ryu, C. M. (2009). Rhizosphere bacteria help plants tolerate abiotic stress. *Trends in Plant Science, 14,* 1–4.

Yanni, Y. G., Rizk, R. Y., Abd El-Fattah, F. K., Squartini, A., Corich, V., Giacomini, A., et al. (2001). The beneficial plant growth-promoting association of *Rhizobium leguminosarum* bv. *trifolii* with rice roots. *Australian Journal Plant of Physiology, 28,* 845–870.

Yedida, I., Srivastava, A. K., Kapulnik, Y., & Chet, I. (2001). Effect of *Trichoderma harzianum* on microelement concentration and increased growth of cucumber plants. *Plant & Soil*, *235*, 235–242.

Zamioudis, C., & Pieterse, C. M. (2012). Modulation of host immunity by beneficial microbes. *Molecular Plant–Microbe Interactions*, *25*, 139–150.

Zhang, J., Liu, J., Meng, L., Ma, Z., Tang, X., Cao, Y., et al. (2012). Isolation and characterization of plant growth-promoting rhizobacteria from wheat roots by wheat germ agglutinin labeled with fluorescein isothiocyanate. *Journal of Microbiology*, *50*, 191–198.

Zhuang, X., Chen, J., Shim, H., & Bai, Z. (2007). New advances in plant growth-promoting rhizobacteria for bioremediation. *Environment International*, *33*, 406–413.

CHAPTER FOUR

Recombinant Production of Spider Silk Proteins

Aniela Heidebrecht*, Thomas Scheibel*,[1]
*Department of Biomaterials, University of Bayreuth, Universitätsstr. 30, 95440 Bayreuth, Germany
[1]Corresponding author: E-mail: thomas.scheibel@bm.uni-bayreuth.de

Contents

Abstract

Natural spider silk fibers combine extraordinary properties such as stability and flexibility which results in a toughness superseding that of all other fiber materials. As the spider's aggressive territorial behavior renders their farming not feasible, the

Advances in Applied Microbiology, Volume 82
ISSN 0065-2164, http://dx.doi.org/10.1016/B978-0-12-407679-2.00004-1

biotechnological production of spider silk proteins (spidroins) is essential in order to investigate and employ them for applications. In order to accomplish this task, two approaches have been tested: firstly, the expression of partial cDNAs, and secondly, the expression of synthetic genes in several host organisms, including bacteria, yeast, plants, insect cells, mammalian cells, and transgenic animals. The experienced problems include genetic instability, limitations of the translational and transcriptional machinery, and low solubility of the produced proteins. Here, an overview of attempts to recombinantly produce spidroins will be given, and advantages and disadvantages of the different approaches and host organisms will be discussed.

1. INTRODUCTION

The outstanding mechanical properties of spider silk fibers have impressed mankind since ancient times (Gerritsen, 2002). Its combination of stability and flexibility gives natural spider silk fibers a toughness no other natural or synthetic fiber can achieve (Gosline, Guerette, Ortlepp, & Savage, 1999). Even compared to Kevlar, one of the strongest known synthetic fibrous materials (Table 4.1), spider silk fibers are superior, because they can absorb three times more energy before breaking (Römer & Scheibel, 2007). In addition, spider silk comprises antimicrobial and hypoallergenic properties suitable in applications, such as in wound dressings. A spider's web has been reported to be able to stop bleeding and to support the healing process by isolating the wound from the surrounding air (Bon, 1710).

The outstanding mechanical properties of spider silk fibers are based on their hierarchical setup and the proteins involved. A spider silk fiber mainly consists of proteins, the so-called spidroins ("spidroin" = spider fibroin)

Table 4.1 The mechanical properties of *A. diadematus* dragline silk compared with other materials

Material	Young's modulus [GPa]	Strength [MPa]	Extensibility [%]	Toughness [MJ/m^3]
A. diadematus dragline	6	700	30	150
B. mori cocoon	7	600	18	70
Elastin	0.001	2	15	2
Nylon 6.6	5	950	18	80
Kevlar 49	130	3600	2.7	50
Steel	200	1500	0.8	6
Carbon fiber	300	4000	1.3	25

Adapted from Gosline et al., (1999); Heim et al., (2009); Madsen et al., (1999).

(Hinman & Lewis, 1992). The spidroin's secondary structure elements provide a combination of stability and flexibility that results in the outstanding toughness of the fiber.

However, in order to both investigate and employ spider silk for applications, a large-scale silk production is necessary. In contrast to farming of the silkworm *Bombyx mori*, farming of spiders is not feasible, because spiders are territorial and show a cannibalistic behavior (Fox, 1975). Additionally, it has been discovered that spiders held in captivity produce less silk and with a lower quality, as its amino acid content directly reflects the spider's diet (Craig et al., 2000). In order to obtain silk proteins on a large scale, the main approach is the recombinant production of natural spidroins or of designed proteins which comprise essential features of natural spidroins. The tested host organisms include bacteria, yeast, as well as insect cells, plants, mammalian cells, and even transgenic mammals. In this review, an overview of recombinantly produced spidroins will be given, and advantages and disadvantages of the different approaches and host organisms will be discussed.

2. SPIDERS AND SPIDER SILK

Web spiders (*Aranea*), the best known order of the *Arachnida*, benefit from their silk every day. Some spiders, such as female orb-weavers, produce up to seven different types of silk and use them for a range of different applications, such as in a complex web for catching prey, for wrapping prey, and for protecting their offsprings. Each silk has unique mechanical properties, depending on their intended use (Stauffer, Coguill, & Lewis, 1994) and is named after the gland in which its proteins are produced (see below).

2.1. The Different Types of Spider Silk and Their Properties

The different types of silk of a female orb web spider have mechanical properties (Table 4.2) adapted to their tasks (Fig. 4.1).

2.1.1. Major Ampullate (MA)/Dragline Silk

Spiders use MA silk with a very high tensile strength (Blackledge, Summers, & Hayashi, 2005; Gosline, Denny, & Demont, 1984; Gosline et al., 1994, 1999, 2002; Thiel & Viney, 1996; Viney, 1997; Vollrath & Porter, 2006) for the outer frame as well as the radii of their web. MA silk also functions as a lifeline for the spider in case it has to escape from a predator (Aprhisiart & Vollrath, 1994).

Table 4.2 Mechanical properties of the different types of spider silk from different spiders

Silk	Young's modulus [GPa]	Strength [MPa]	Extensibility [%]	Toughness [MJ/m^3]	Reference
MA					
Argiope trifasciata	9.3	1290	22	145	(Hayashi et al., 2004)
MI					
Argiope trifasciata	8.5	342	54	148	(Hayashi et al., 2004)
Flagelliform					
Araneus diadematus	0.003	500	270	150	(Gosline et al., 1999)
Cylindriform					
Argiope bruennichi	9.1	390	40	128	(Zhao et al., 2006)
Aciniform					
Argiope trifasciata	9.6	687	86	367	(Hayashi et al., 2004)

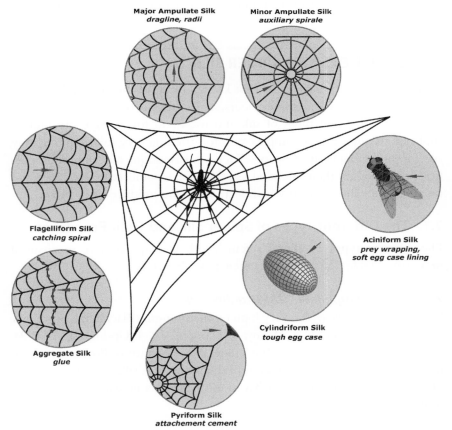

Figure 4.1 Schematic presentation of the seven different types of spider silk. (For color version of this figure, the reader is referred to the online version of this book.)

The core of the MA silk fiber consists of fibrils comprising at least two MA spidroins, with a molecular weight of 200–350 kDa (Ayoub, Garb, Tinghitella, Collin, & Hayashi, 2007; Candelas & Cintron, 1981; Jackson & O'Brien, 1995; O'Brien, Fahnestock, Termonia, & Gardner, 1998). In principle, MA spidroins can be divided into two classes, one poor in proline residues (MaSp1) and one proline-rich (MaSp2). All MA spidroins consist of repetitive core sequences that are flanked by nonrepetitive, folded amino- and carboxy-terminal domains. Individual amino acid motifs of the core sequences (see below) are repeated up to 100 times (Ayoub et al., 2007; Hinman & Lewis, 1992), and generally a motif contains 40–200 amino acids (Ayoub et al., 2007; Guerette, Ginzinger, Weber, & Gosline, 1996; Hayashi, Blackledge, & Lewis, 2004).

Strikingly, the proline content varies greatly between spider species, indicating a different ratio of MaSp1 and MaSp2 (Andersen, 1970; Lombardi & Kaplan, 1990; Mello, Senecal, Yeung, Vouros, & Kaplan, 1994). Due to its molecular structure, a pyrrolidine ring, a proline residue inflicts steric constraints on the protein backbone, since it doesn't supply an amide proton that can take part in hydrogen bonds (Hurley, Mason, & Matthews, 1992; Vollrath & Porter, 2006). Thus, proline residues favor the formation of β-turn and γ-turn structures over α-helix and β-sheet structures (Hayashi, Shipley, & Lewis, 1999; Thiel, Guess, & Viney, 1997; Zhou, Wu, & Conticello, 2001). The proline-containing structured motifs, such as GPGXX, are not capable of forming crystalline structures, which result in a higher elasticity of the fiber (Hayashi et al., 1999; Thiel et al., 1997). The supercontraction of a fiber, which is closely linked to its mechanical properties, is also based on its proline content (Liu, Shao, & Vollrath, 2005; Liu, Shao, et al., 2008). Silkworm and spider MI silk both contain only a small amount of proline and show almost no supercontraction (Colgin & Lewis, 1998; Ito et al., 1995; Jelinski et al., 1999; Vollrath, 1994), whereas the supercontraction of MA silks differs from species to species, since their proline content differs as well (Vollrath & Porter, 2006). An intrinsic disorder in combination with hydration, as it is caused by proline- and glycine residues, is fundamental for an elastomeric function of silk fibers (Rauscher, Baud, Miao, Keeley, & Pomes, 2006).

Strikingly, in the inner core of the MA silk fiber of *Nephila clavipes*, both MaSp1 and MaSp2 were found (Sponner et al., 2007), while the surrounding area contains only MaSp1, whose proline content is low.

The repetitive domains of MA spidroins of *Araneus diadematus* and *N. clavipes* contain stretches of polyalanine, as well as motifs containing $(GGX)_n$ or GPGXX (X = tyrosine, leucine, glutamine) (Winkler & Kaplan, 2000). The alanine-rich areas form crystalline β-sheets (Thiel & Viney, 1996) and grant the high tensile strength of the spider silk fibers (Simmons, Michal, & Jelinski, 1996). In contrast, the $(GGX)_n$ and GPGQQ (ADF3) or GPGXX (ADF4) sequences fold into 3_1-helices (Hijirida et al., 1996) and β-turn spirals, respectively. These glycine-rich areas provide an amorphous matrix for the crystalline β-sheets (Perez-Rigueiro, Elices, Plaza, & Guinea, 2007; Sponner et al., 2007; Sponner, Unger, et al., 2005), and thus generate the elasticity and flexibility of the spider silk fiber as mentioned above (Brooks, Steinkraus, Nelson, & Lewis, 2005; Liu, Sponner, et al., 2008; Ohgo, Kawase, Ashida, & Asakura, 2006). The structural motifs of MA spidroins, as well as Minor ampullate (MI) spidroins and Flagelliform silk are shown in Fig. 4.2.

The nonrepetitive terminal domains comprise 100–150 amino acids and have α-helical secondary structures arranged in a five-helix bundle (Challis, Goodacre, & Hewitt, 2006; Hedhammar et al., 2008; Huemmerich, Helsen, et al., 2004; Rising, Hjalm, Engstrom, & Johansson, 2006). These terminal

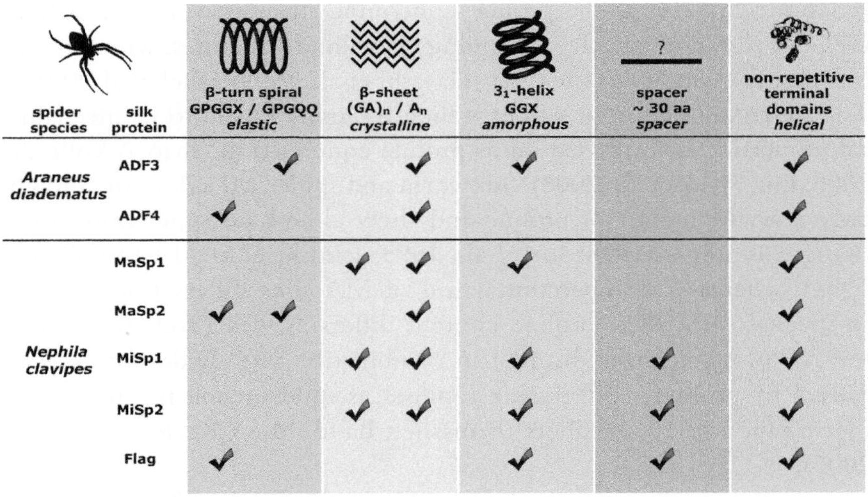

spider species	silk protein	β-turn spiral GPGGX / GPGQQ elastic	β-sheet $(GA)_n$ / A_n crystalline	3_1-helix GGX amorphous	spacer ~30 aa spacer	non-repetitive terminal domains helical
Araneus diadematus	ADF3	✓	✓	✓		✓
	ADF4	✓	✓			✓
Nephila clavipes	MaSp1		✓	✓		✓
	MaSp2	✓	✓			✓
	MiSp1		✓	✓	✓	✓
	MiSp2	✓	✓	✓	✓	✓
	Flag	✓		✓	✓	✓

Figure 4.2 Structural motifs of various spider silk proteins from *A. diadematus* and *N. clavipes*. X: predominantly tyrosine, leucine, glutamine, alanine, and serine. aa = amino acid. (For color version of this figure, the reader is referred to the online version of this book.)

domains enable the storage of spidroins at high concentrations in the spinning duct and play an important role during the initiation of fiber assembly (Askarieh et al., 2010; Eisoldt, Hardy, Heim, & Scheibel, 2010; Eisoldt, Scheibel, & Smith, 2011; Eisoldt, Thamm, & Scheibel, 2012; Hagn, Thamm, et al., 2010).

2.1.2. MI Silk

MI silk, which has similar properties as MA silk, is used by the spider as an auxiliary spiral to stabilize the scaffold during web construction (Dicko, Knight, Kenney, & Vollrath, 2004; Riekel & Vollrath, 2001; Tillinghast & Townley, 1994). MI silks are mainly composed of two proteins, MiSp1 and MiSp2, which have molecular weights of approx. 250 kDa (Candelas & Cintron, 1981), and strikingly differ considerably from MA silk in their composition, as they contain almost no proline residues (Fig. 4.2). Further, their glutamic acid content is significantly reduced (Andersen, 1970). MiSp1 and MiSp2 of *N. clavipes* contain repetitive sequences that are composed of 10 repeat units. In the case of MiSp1, one repeat unit contains two motifs, GGXGGY (X = glutamine or alanine) alternating with $(GA)_y(A)_z$ (y = 3–6, z = 2–5) (Colgin & Lewis, 1998). MiSp2 of *N. clavipes* comprises $(GGX)_n$ (X = tyrosine, glutamine or alanine, n = 1–3) blocks, alternating with GAGA. In contrast to MA spidroins, the repetitive regions of MI spidroins are disrupted by 137 amino acid-long nonrepetitive serine-rich spacer regions being almost identical in MiSp1 and MiSp2. Comparable to MA spidroins, crystalline structures are embedded in an amorphous matrix (Colgin & Lewis, 1998). However, NMR studies showed that in contrast to MA silk, only a small fraction of alanine residues contribute to crystalline β-sheets in MI silk. Thus, the high tensile strength of MI silk cannot solely be due to β-sheet structures, but must also be based on other structural features, and in this content it is assumed that cross-linking as well as specific matrix properties of MI proteins, different to those of MA proteins, have some impact thereon (Dicko et al., 2004).

2.1.3. Flagelliform Silk

Flagelliform silk is highly elastic and is used as the capture spiral of an orb web, because it can absorb the high kinetic energy that results from the impact of an insect (Becker et al., 2003; Bini, Knight, & Kaplan, 2004; Brooks et al., 2005; Dicko et al., 2004; Ohgo et al., 2006; Scheibel, 2004; Winkler & Kaplan, 2000). Flagelliform silk, which is also called "viscid" silk, mainly consists of one protein that has a molecular weight of approx. 500 kDa and contains more proline and valine residues, whereas its alanine content is reduced in comparison to MA and MI silks (Andersen, 1970). The *N. clavipes* flagelliform protein comprises blocks of $(GGX)_n$ and GPGXX, which build 3_1-helices and β-turn spirals, respectively. These secondary structural motifs are responsible for the high elasticity and flexibility of this type of silk (Brooks et al., 2005; Liu, Sponner, et al., 2008; Ohgo et al., 2006). $(GPGGX)_2$ motifs (X = serine or tyrosine) form β-turns, and assemble a structure similar to a spring (Hayashi et al., 1999). Since flagelliform silk has more than 40 adjacent linked β-turns in spring–like spirals, it is likely that this structure adds to the extraordinary elasticity (200%) of the fiber (Hayashi & Lewis, 1998).

2.1.4. Pyriform Silk

Pyriform silk is used by the spider to securely attach individual MA, MI, and Flagelliform silk fibers to each other as well as to a substrate, such as a tree branch or a wall (Hajer & Rehakova, 2003; Kovoor & Zylberberg, 1980). Pyriform silk proteins of *A. diadematus* have a randomly coiled structure as they contain a low amount of small nonpolar amino acids as well as significant quantities of polar and charged amino acids, which are important for physical cross-linking (Andersen, 1970).

2.1.5. Aggregate Silk

In order to prevent prey to escape the spider's web, ecribellate spiders, such as *Latrodectus hesperus*, cover their capture spiral with an aggregate silk, which is a mixture of sticky glycoproteins and small highly hygroscopic peptides (Hu et al., 2007; Vollrath, 2006; Vollrath & Tillinghast, 1991). The aggregate silk proteins of *L. hesperus* (Hu et al., 2007) contain a low amount of small nonpolar amino acids, as well as significant amounts of proline residues, polar and charged amino acids (Andersen, 1970). In contrast to that, cribellate spiders, e.g. from the *Uloborus* sp., surround their capture spiral not with an aqueous sticky glue but with thin cribellar fibrils, which are 10 nm in diameter. The stickiness of these dry cribellar fibrils is accomplished through a combination of van der Waals and hygroscopic forces (Hawthorn & Opell, 2002, 2003).

2.1.6. Cylindriform/Tubiliform Silk

Araneus diadematus cylindriform silk proteins are composed of polyalanine blocks in combination with $(GGX)_n$ (X = alanine, leucine, glutamine, or tyrosine) motifs, which form β-sheet stacks (similar to *N. clavipes* MA silk) (Barghout, Thiel, & Viney, 1999), giving the cylindriform fibers a high strength. They are suitable to be used by the spider as a tough egg case in order to protect its offspring (Bittencourt et al., 2007; Garb & Hayashi, 2005; Huang et al., 2006; Hu, Kohler, et al., 2005, 2006; Hu, Lawrence, et al., 2005; Hu, Vasanthavada, et al., 2006; Kohler et al., 2005; Rising et al., 2006).

2.1.7. Aciniform Silk

Female orb web spiders use aciniform silk for multiple purposes, such as a soft lining inside the egg case, as well as in order to reinforce the prey wrapping and the pyriform silk attachment cement (Blackledge & Hayashi, 2006; Hayashi et al., 2004; La Mattina et al., 2008; Vasanthavada et al., 2007). Even though the 300 kDa AcSp (*L. hesperus* (Vasanthavada et al., 2007)) contains $(GGX)_n$ domains, they differ greatly from other types of silk proteins (Andersen, 1970).

2.2. Hierarchical Structure of MA Silk Fibers

MA threads of the orb-weaver *N. clavipes* show a core–shell structure (Fig. 4.3). Its proteinaceous core is composed of MA spidroin fibrils, which are oriented in direction of the fiber axis, and which itself can be divided into an inner and outer region, based on the protein content. The core is covered by a 150–250 nm–thick three-layer shell (Augsten, Muhlig, & Herrmann, 2000; Augsten, Weisshart, Sponner, & Unger, 1999), consisting of MI silk, glycoproteins, and lipids (Sponner et al., 2007).

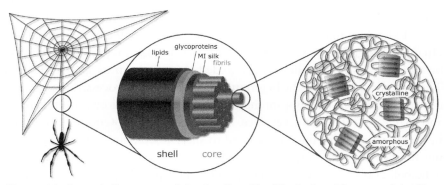

Figure 4.3 *Core–shell structure of the dragline silk of N. clavipes.* The core of the fiber comprises fibrils that are oriented along the fiber axis. On a molecular level, these fibrils consist of crystalline areas that are embedded in an amorphous matrix, depending on the amino acid composition. The core is covered by a three-layer shell containing MI silk, glycoproteins, and lipids. (For color version of this figure, the reader is referred to the online version of this book.)

A 10–20 nm-thick lipid layer builds the outer coat of the fiber (Sponner et al., 2007), and it fulfills a number of tasks, as it serves as a carrier for pheromones that play an important role in the sex and species recognition (Schulz, 2001). As this coat is just loosely attached, it offers only limited protection from the surrounding environment and does not contribute to the mechanical properties of the fiber (Sponner et al., 2007). The under-lying layer is about 40–100 nm thick and composed of glycoproteins (Augsten et al., 2000). This layer offers a more profound protection from microorganisms, since it is tighter attached than the lipid layer. Additionally, the glycoprotein coat indirectly influences the mechanical strength of the fiber, as it is able to regulate the water balance, which has an impact on the contraction state of the fiber (Liu et al., 2005). Even though the proteins in the glycoprotein layer differ from those in the skin and the core of the fiber, their molecular weights are comparable (Sponner et al., 2007). The inner layer of the fiber's shell is 50–100 nm thick and comprises MI silk proteins. This layer plays a dual role: firstly, it protects the fiber from environmental damage inflicted by chemical agents and microbial activity, and, secondly, it mechanically supports the fiber as it adds plasticity (Sponner et al., 2007).

2.3. Natural MA Spidroin Processing

The spider's silk glands contain a highly concentrated protein solution (up to 50% (w/v) (Hijirida et al., 1996)), the so-called spinning dope. In the spinning gland, the protein solution is stored at a neutral pH and high salt concentration that prevents aggregation of the proteins. It is assumed that the proteins form supramolecular micelle-like structures (Eisoldt et al., 2010; Exler, Hummerich, & Scheibel, 2007; Hagn, Eisoldt, et al., 2010; Lin, Huang, Zhang, Fan, & Yang, 2009), with the unfolded repeated sequences being inside of the micelle and the folded polar ends on its surface (Hagn, Thamm, et al., 2010). At the beginning of the assembly process, the spinning dope is transferred to the spinning duct, where acidification, ion exchange, and extraction of water take place. Additionally, the proteins are arranged in the direction of strain and exposed to shear forces, due to the accelerat-ing flow of the spinning dope (Hardy & Scheibel, 2009; Heim, Keerl, & Scheibel, 2009; Knight & Vollrath, 1999). The pH drop from 7.2 in the gland to 6.3 in the spinning duct causes dimerization of the amino-terminal domains of the proteins into antiparallel dimers (head-to-tail) (Askarieh et al., 2010; Eisoldt et al., 2010; Hagn, Thamm, et al., 2010). The lowering of the sodium chloride concentration (Knight & Vollrath, 2001; Tillinghast, Chase, & Townley, 1984) in combination with shear forces results in the alignment of the protein chains parallel to the fiber axis (Eisoldt et al., 2010; van Beek, Hess, Vollrath, & Meier, 2002). During this assembly step, the carboxy-terminal domains control the correct orientation of the repetitive regions, and thus support the formation of β-sheet stacks, which strongly influence the mechanical properties of the spider silk fiber (Hagn, Eisoldt, et al., 2010). After being pulled from the spinneret, the excess water of the fibers evaporates, and fiber formation is completed.

2.4. Artificial Fiber Formation

In order to use the outstanding mechanical properties of spider silk fibers for industrial purposes, many attempts have been made to spin reconstituted silk into fibers. For this purpose, natural silk fibers were dissolved in different solvents, and the artificial silk dopes were spun with several techniques, ranging from wet-spinning into alcohol baths (Matsumoto, Uejima, Iwasaki, Sano, & Sumino, 1996) (fiber formation by protein precipitation in a coagulation bath) to electrospinning (Zarkoob et al., 2004) (solvent evaporation by accelerating the spinning solution in an electrical field). The reconstituted silk fibers did neither have the mechanical properties (Lazaris et al., 2002; Liivak, Blye, Shah, & Jelinski, 1998; Madsen, Shao, & Vollrath, 1999; Marsano et al., 2005; Shao, Vollrath, Yang, & Thogersen, 2003; Xie, Zhang, Shao, & Hu, 2006; Yao, Masuda, Zhao, & Asakura, 2002) nor the structural integrity (Putthanarat, Stribeck, Fossey, Eby, & Adams, 2000) of natural silk fibers.

Further, the properties of natural and reconstituted silk dopes showed significant differences (Holland, Terry, Porter, & Vollrath, 2007). Due to the harsh conditions that are necessary to dissolve the silk fibers for dope preparation, high temperatures and chaotropic agents cause severe degradation of the silk proteins. It has been shown that the molecular weight of the proteins (Iridag & Kazanci, 2006; Yamada, Nakao, Takasu, & Tsubouchi, 2001; Zuo, Dai, & Wu, 2006) as well as their conformation (Asakura, Kuzuhara, Tabeta, & Saito, 1985; Zuo et al., 2006) is significantly altered during the silk dope preparation.

3. RECOMBINANT PRODUCTION OF SPIDER SILK PROTEINS

As mentioned above, the livestock breeding and harvesting the silk fibers, like it is done with the mulberry silkworm *B. mori* (Yamada et al., 2001), is not possible with spiders. In contrast to *B. mori*, the spider's aggressive territorial behavior and its cannibalism render farming not feasible in large scale, hampering the accessibility to the raw material. Furthermore, spiders that are held in captivity produce silk with lower qualities, likely depending on their nutrition (Madsen et al., 1999; Vollrath, 1999). In order to overcome these difficulties, different approaches have been tested to produce recombinant spider silk proteins in large quantities, followed by artificial processing similar to the attempts with regenerated silk proteins.

One tested route is the expression of natural spider silk genes in different host organisms. Apart from prokaryotes and eukaryotes, even transgenic animals have been used as host organisms with varying results. An overview of the recombinantly produced proteins based on native silk genes in different host organisms is given in Table 4.3.

Limited success has been achieved by expressing partial cDNAs in different host organisms, as differences in codon usage cause inefficient translations. Additionally, the highly repetitive nature of spider silk genes impedes their manipulation and amplification using polymerase chain reaction (Heim et al., 2009; Scheibel, 2004).

In order to overcome these problems, synthetic genes were produced encoding proteins that differ from the natural spider silk proteins, but possess their key features. These synthetic mimics of spider silk genes were created by identifying relevant amino-acid sequence motifs of natural spidroins (see above) and subsequently back-translating these into DNA sequences considering the codon usage of the corresponding host organism (Huemmerich, Helsen, et al., 2004; Lewis, Hinman, Kothakota, & Fournier, 1996; Prince, Mcgrath, Digirolamo, & Kaplan, 1995). An overview of recombinantly produced spider silk proteins in bacterial host organisms using synthetic genes is given in Table 4.4, and in eukaryotic hosts in Table 4.5.

3.1. Bacterial Expression Hosts Used for Spider Silk Production

3.1.1. *Escherichia coli*

Escherichia coli is a gram-negative enterobacterium, which is often used as a host organism for recombinant protein production. Its relative simplicity, its well-known genetics, and the capability of fast high-density cultivation render it a suitable host organism for the fast and inexpensive large-scale production of recombinant proteins. The availability of different plasmids, fusion protein partners, and mutated strains are additional advantages (Sorensen & Mortensen, 2005).

Table 4.3 Overview of natural spider silk genes expressed in different host organisms

Type of silk protein	Terminal domains*	M$_W$** [kDa]	Yield** [mg/L]	Spider	Host organism	References
MA silk						
MaSp1	–	22, 25	N/A	*Euprosthenops* sp.	Mammalian cells (COS–1)	(Grip et al., 2006)
	±NTD, ±CTD	10–28	25–150	*Euprosthenops australis*	Bacteria (*Escherichia coli*)	(Askarieh et al., 2010; Hedhammar et al., 2008; Stark et al., 2007)
	+CTD	43	4	*Nephila clavipes*	Bacteria (*Escherichia coli*)	(Arcidiacono et al., 1998)
MaSp1 (fibroin chimera)	–	83	N/A	*Nephila clavata*	Transgenic animals (*Bombyx mori*)	(Wen et al., 2010)
MaSp1 & MaSp2	±CTD	12	N/A	*Nephila clavipes*	Bacteria (*Escherichia coli*)	(Sponner, Vater, et al., 2005)
	+CTD	60–140	N/A	*Nephila clavipes*	Mammalian cells (MAC–T & BHK)†	(Lazaris et al., 2002)
	–	31–66	8–12	*Nephila clavipes*	Transgenic animals (mice)	(Xu et al., 2007)
	–	33–39	N/A	*Nephila clavipes*	Yeast (*Pichia pastoris*)	(Teule, Aube, Ellison, & Abbott, 2003)
MaSp1- & MaSp2- collagen fusion protein	–	57–61	N/A	*Nephila clavipes*	Yeast (*Pichia pastoris*)	(Teule et al., 2003)

Continued

Table 4.3 Overview of natural spider silk genes expressed in different host organisms—cont'd

Type of silk protein	Terminal domains*	M_W** [kDa]	Yield** [mg/L]	Spider	Host organism	References
ADF3	+CTD	60–140	25–50	*Araneus diadematus*	Mammalian cells (MAC-T & BHK†)	(Lazaris et al., 2002)
	–	60	N/A	*Araneus diadematus*	Transgenic animals (goats)	(Karatzas et al., 1999)
ADF3 & ADF4	±CTD	50–105	50	*Araneus diadematus*	Insect cells (*Spodoptera frugiperda*)	(Ittah et al., 2006)
	+CTD	35–56	N/A	*Araneus diadematus*	Insect cells (*Spodoptera frugiperda*)	(Huemmerich, Scheibel, et al., 2004)
Flagelliform						
Flag	+CTD	28	N/A	*Araneus ventricosus*	Insect cells (*Spodoptera frugiperda*)	(Lee et al., 2007)
Polyhedron-Flag fusion protein	+CTD	61	N/A	*Araneus ventricosus*	Insect cells (*Spodoptera frugiperda*)	(Lee et al., 2007)
Tubiliform						
TuSp1	±NTD, ±CTD	12–15	N/A	*Nephila antipodiana*	Bacteria (*Escherichia coli*)	(Lin et al., 2009)
TuSp1	±CTD	33, 45	4.8–7.2	*Latrodectus hesperus*	Bacteria (*Escherichia coli*)	(Gnesa et al., 2012)
Pyriform						
PySp2	±CTD	N/A	N/A	*Latrodectus hesperus*	Bacteria (*Escherichia coli*)	(Geurts et al., 2010)

*NTD: amino-terminal domain, CTD: carboxy-terminal domain.
**N/A: not applicable.
†MAC-T: bovine mammary epithelial alveolar cells, BHK: baby hamster kidney cells.

Even though several work groups succeeded in producing recombinant spider silk-like proteins using *E. coli* as a host organism (Arcidiacono, Mello, Kaplan, Cheley, & Bayley, 1998, 2002; Bini et al., 2006; Brooks, Stricker, et al., 2008; Cappello & Crissman, 1990; Fukushima, 1998; Huang, Wong, George, & Kaplan, 2007; Huemmerich, Slotta, & Scheibel, 2006; Slotta, Rammensee, Gorb, Scheibel, 2008; Slotta, Tammer, Kremer, Koelsch, & Scheibel, 2006; Stephens et al., 2005; Winkler, Wilson, & Kaplan, 2000; Zhou et al., 2001), limitations such as low protein yields (Arcidiacono et al., 1998; Prince et al., 1995), heterologous proteins (Fahnestock & Irwin, 1997; Lazaris et al., 2002; Xia et al., 2010), and low protein solubility (Bini et al., 2006; Fukushima, 1998; Mello, Soares, Arcidiacono, & Butlers, 2004; Szela et al., 2000; Winkler et al., 2000; Wong Po Foo et al., 2006) have been encountered. Translation limitations are caused by unwanted recombination events shortening the repetitive genes (Arcidiacono et al., 1998; Fahnestock & Irwin, 1997; Rising, Widhe, Johansson, & Hedhammar, 2011) and depletion of tRNA pools due to the guanine- and cytosine-rich genes (Rosenberg, Goldman, Dunn, Studier, & Zubay, 1993). Moreover, an overproduction of recombinant spider silk proteins in *E. coli* can lead to the formation of inclusion bodies (Liebmann, Huemmerich, Scheibel, & Fehr, 2008), which comprise incorrect or incompletely folded proteins and which retard cell growth (Sorensen & Mortensen, 2005). Fairly good protein yields have been obtained for recombinant spider silk proteins with a molecular weight up to approx. 100 kDa. With an increasing molecular weight of the recombinant spider silk proteins, the impact of these limitations increases greatly.

Escherichia coli is incapable of performing most eukaryotic posttranslational modifications (PTMs), such as glycosylation and phosphorylation of proteins. The dragline silk of *N. clavipes* contains phosphorylated tyrosine and serine residues (Michal, Simmons, Chew, Zax, & Jelinski, 1996). Since the biological function and the correct folding of proteins are dependent on PTMs (Gellissen et al., 1992; Kukuruzinska & Lennon, 1998), it is suspected that the phosphorylated residues influence the processing of the spidroins into fibers, even though the impact on the physical properties has not yet been determined (Michal et al., 1996). However, as nowadays recombinant spidroins have successfully been produced in *E. coli*, the PTMs do not seem crucial at least for the protein production.

A native-sized recombinant spider silk protein (285 kDa) was produced using metabolically engineered *E. coli* (Xia et al., 2010). By analyzing the influence of the silk gene expression on the host protein synthesis, it was found that the silk gene expression causes stress to the host cells, since

Table 4.4 Overview of synthetic spider silk genes expressed in bacterial host organisms

Type of silk protein**	Terminal domains*	M_W** [kDa]	Yield** [mg/L]	Spider**	Host organism	References
MA silk						
MaSp1	–	N/A	N/A	*Latrodectus hesperus*	*Salmonella typhimurium*	(Widmaier & Voigt, 2010; Widmaier et al., 2009)
	–	100–285	1.2 g/L	*Nephila clavipes*	*Escherichia coli*[†]	(Xia et al., 2010)
	–	15–26	16–95	*Nephila clavipes*	*Escherichia coli*	(Szela et al., 2000; Winkler et al., 1999, 2000)
	–	45–60	2.5–150	*Nephila clavipes*	*Escherichia coli*	(Bini et al., 2006; Huang et al., 2007; Wong Po Foo et al., 2006)
MaSp1 (Gly-rich repeats)	–	10–20	1.2–5.2	*Nephila clavipes*	*Escherichia coli*	(Fukushima, 1998)
MaSp2	–	63–71	N/A	*Argiope aurantia*	*Escherichia coli*	(Brooks, Stricker, et al., 2008; Teule et al., 2009)
	–	31–112	N/A	N/A	*Escherichia coli*	(Lewis et al., 1996)
MaSp2/Flag	–	58, 62	7–10	*Nephila clavipes*	*Escherichia coli*	(Teule, Furin, Cooper, Duncan, & Lewis, 2007)
MaSp1 & MaSp2	NTD	14	N/A	*Latrodectus hesperus*	*Escherichia coli*	(Hagn, Thamm, et al., 2010)
	±CTD	20–56	15–35	*Nephila clavipes*	*Escherichia coli*	(Arcidiacono et al., 2002; Mello et al., 2004)
	–	N/A	N/A	*Nephila clavipes*	*Bacillus subtilis*	(Fahnestock, 1994)
	–	55, 67	N/A	*Nephila clavipes*	*Escherichia coli*	(Brooks, Nelson, et al., 2008)
	–	15–41	2–15	*Nephila clavipes*	*Escherichia coli*	(Prince et al., 1995)
	–	65–163	N/A	*Nephila clavipes*	*Escherichia coli*	(Fahnestock & Irwin, 1997)

ADF3, ADF4	±CTD	34–106	N/A	Araneus diadematus	Escherichia coli	(Huemmerich, Helsen, et al., 2004; Schmidt, Romer, Strehle, & Scheibel, 2007)
ADF1, ADF2, ADF3, ADF4	–	25–56	N/A	Araneus diadematus	Salmonella typhimurium	(Widmaier & Voigt, 2010; Widmaier et al., 2009)
Flagelliform silk						
Flag	–	N/A	N/A	Nephila clavipes	Salmonella typhimurium	(Widmaier & Voigt, 2010; Widmaier et al., 2009)
Flag	±CTD	14–94	N/A	Nephila clavipes	Escherichia coli	(Heim, Ackerschott, & Scheibel, 2010)
Flag (Gly-rich repeats)	–	25	11.6	Nephila clavipes	Escherichia coli	(Zhou et al., 2001)
N/A						
N/A	–	76–89	N/A	N/A	Escherichia coli	(Cappello et al., 1990)
Gly-rich repeats & Ala-blocks	–	18–36	21–41	N/A	Escherichia coli	(Yang & Asakura, 2005)

*NTD: amino-terminal domain, CTD: carboxy-terminal domain.
**N/A: not applicable.
†Metabolically engineered.

Table 4.5 Overview of synthetic spider silk genes expressed in different eukaryotic host organisms

Type of silk protein**	Terminal* domains	M_W** [kDa]	Yield** [mg/L]	Spider**	Host organism	References
MA silk						
MaSp1	–	N/A	~300	*Nephila clavipes*	Yeast (*Pichia pastoris*)	(Fahnestock & Bedzyk, 1997)
	–	94	N/A	*Nephila clavipes*	Yeast (*Pichia pastoris*)	(Agapov et al., 2009; Bogush et al., 2009)
	–	64, 127	N/A	*Nephila clavipes*	Plants (*Arabidopsis thaliana*)	(Barr et al., 2004; Yang et al., 2005)
	–	13–100	80	*Nephila clavipes*	Plants (Tobacco: *Nicotiana tabacum*)	(Scheller & Conrad, 2005; Scheller et al., 2001, 2004)
	–	13–100	80	*Nephila clavipes*	Plants (Potato: *Solanum tuberosum*)	(Scheller & Conrad, 2005; Scheller et al., 2001, 2004)
	–	70	N/A	*Nephila clavata*	Insect cells (*Bombyx mori*)	(Zhang et al., 2008)
	–	70	6[†]	*Nephila clavata*	Transgenic animals (*Bombyx mori*)	(Zhang et al., 2008)
MaSp2	–	113	N/A	*Nephila madagascariensis*	Yeast (*Pichia pastoris*)	(Bogush et al., 2009)
	+CTD	65	N/A	*Nephila clavipes*	Plants (Tobacco: *Nicotiana tabacum*)	(Patel et al., 2007)

MaSp1 & MaSp2	+CTD	60	0.3–3‡	*Nephila clavipes*	Plants (Tobacco: *Nicotiana tabacum*)	(Menassa et al., 2004)
	–	50	N/A	*Nephila clavipes*	Transgenic animals (goat)	(Perez-Rigueiro et al., 2011)
Flagelliform						
Flag	–	37	13.3	*Nephila clavipes*	Insect cells (*Bombyx mori*)	(Miao et al., 2006)
N/A						
Amphiphilic silk-like protein	–	28–32	1–3 g/L	N/A	Yeast (*Pichia pastoris*)	(Werten et al., 2008)

*NTD: amino-terminal domain, CTD: carboxy-terminal domain.
**N/A: not applicable.
†6 mg/larva.
‡0.3–3 mg/kg of tissue.

many stress-response proteins were upregulated. In order to increase the yield of large recombinant glycine-rich proteins (100–285 kDa), the over-expression of glycyl-tRNA synthetase, as well as the elevation of metabolic pools of tRNA^{Gly} and glycine, was tested. By elevating the tRNA^{Gly} pool, an increased production of the 147–285 kDa proteins as well as an enhanced cell growth by 30–50% was observed (Xia et al., 2010). Additionally, amongst addition of glycine to the expression medium, the inactivation of the glycine cleavage system and a glyA overexpression only the latter increased the yield of the largest proteins 10- to 35-fold (Xia et al., 2010). The produced proteins were purified by acidic precipitation of the cell lysate, followed by a fractional ammonium sulfate precipitation. After solubilizing the precipitated protein, dialysis and freeze drying, 1.2 g protein with a purity of ~90% was obtained from 1 L of cell suspension from a high cell-density cultivation (Xia et al., 2010). The purified proteins were dissolved in hexafluoroisopropanol (HFIP), and a 20% (w/v) solution was spun into fibers. The obtained fibers showed a tenacity (508 ± 108 MPa), elongation (15 ± 5%) and Young's modulus (21 ± 4 GPa) similar to those of *N. clavipes* dragline silk fibers (740–1200 MPa; 18–27% and 11–14 GPa, respectively) (Xia et al., 2010). The attempt to spin fibers from a recombinant 377 kDa protein using the stated process failed, because not only the target protein but also several truncated forms were obtained after purification. It is assumed that this is caused by a limitation of the *E. coli* translational machinery (Xia et al., 2010).

3.1.2. *Salmonella typhimurium*

Apart from *E. coli*, other bacterial hosts have been tested for the recombinant silk production, as codon usage of different host organisms can be compared with increasing genomic's knowledge (Terpe, 2006). Other bacterial hosts, such as the gram-negative bacteria *S. typhimurium*, have some advantages over *E. coli*, since *S. typhimurium* has the capability to secrete proteins, which simplifies the isolation of the target silk protein from foreign proteins. For the large-scale production of recombinant spider silk proteins, secretion of the target proteins into the extracellular environment seems to be the feasible way, as no production limitations will be reached due to the lack of intracellular space of the host cells. In general, the secretion of recombinant proteins into the extracellular environment of gram-negative bacteria is difficult, as the proteins have to transfer through an inner and outer membrane. Furthermore, due to its complexity, the underlying mechanism has not yet been fully understood (Harvey et al., 2004; Lee, Tullman-Ercek, & Georgiou, 2006).

Using *Salmonella* SPI-1 T3SS (Type III secretion system encoded on *Salmonella* Pathogenicity Island I), the secretion of recombinant proteins into the extracellular environment of the cells has been shown. SPI-1 T3SS forms needle-like structures that cross both the inner and outer membrane and allow secretion (Marlovits et al., 2006). The mechanism of the protein secretion through the needle into the extracellular environment is highly complex. It requires the co-expression of an amino-terminal peptide TAG, which is not cleaved after secretion, as well as a corresponding chaperone that targets the proteins to the SPI-1 T3SS and a chaperone-binding domain (CBD) (Galan & Collmer, 1999). In order to be able to cleave the amino-terminal TAG after the protein production, an additional cleavage site has to be included within the protein. In order to be able to transfer through the needle, the proteins must at least be partially unfolded due to the dimensions of the needle and have to refold outside of the cells (Widmaier et al., 2009). The amino-terminal TAG, the co-expression of the chaperones and CBD and the partial folding of the target proteins, complicates subsequent purification processes.

Based on the *A. diadematus* proteins ADF1, ADF2, and ADF3, synthetic genes were created that encode proteins (25–56 kDa), which were secreted to the extracellular environment (Widmaier et al., 2009). A tobacco etch virus (TEV) protease site was included following the amino-terminal TAG. Using different TAG/chaperone pairs in combination with the different recombinant proteins, the amount of totally secreted protein ranged between 0.7 and 14 mg/L after 8 h (Widmaier et al., 2009). A secretion efficiency of 7–14%, defined as the ratio of the secreted to the expressed protein, was obtained (Widmaier et al., 2009).

In an additional work, synthetic spider silk genes coding for fragments of *B. mori* cocoon silk proteins, *N. clavipes* flagelliform silk protein, *A. diadematus* ADF1, ADF2, ADF3 and ADF4 and *L. hesperus* MaSp1 were expressed using SPI-1 T3SS (Widmaier & Voigt, 2010). Whereas many of the tested proteins were produced, only a few were secreted into the extracellular environment. The degree of protein secretion was found to be dependent on the length of the recombinant silk protein. All proteins with less than 628 amino acids (including amino-terminal TAG and TEV-recognition site) were secreted, whereas only a fraction of proteins with up to 863 amino acids and no proteins over 863 amino acids were secreted. (Widmaier & Voigt, 2010).

3.1.3. Bacillus subtilis
The gram-positive bacteria *B. subtilis* was used for the recombinant production of spidroins, as it is able to secrete large quantities of the target protein

into the medium, but its secretion system is much simpler than those of yeasts and filamentous fungi (Fahnestock, Yao, & Bedzyk, 2000).

The production of consensus sequences of the repeating units from MaSp1 and MaSp2 of *N. clavipes* in this host organism has been published in a patent (Fahnestock, 1994), but information concerning the produced proteins such as the molecular weight, the obtained yield, and purity has not been published so far.

3.2. Yeast

The yeast *Pichia pastoris* is a promising alternative host organism to *E. coli*, because it is able to correctly process high molecular-weight proteins (Cregg, 2007). Proteins that could not be produced efficiently in bacteria, *Saccharomyces cerevisiae*, or other host organisms, have been successfully produced in *P. pastoris* (Cereghino, Cereghino, Ilgen, & Cregg, 2002). *Pichia pastoris* is a methylotrophic yeast that can use methanol as the sole carbon and energy source. Additionally, this organism is able to grow to high cell densities (>100 g/L dry cell mass) (Cereghino et al., 2002). Importantly, *P. pastoris* allows cytosolic as well as secretory protein production (Macauley-Patrick, Fazenda, McNeil, & Harvey, 2005).

Even upon the production of silk proteins in *P. pastoris*, a variety of protein sizes were obtained (Fahnestock & Bedzyk, 1997). This is due to *P. pastoris'* ability to integrate multiple gene copies into its genome. Because of the highly repetitive sequence of the recombinant spider silk genes, it is likely that the second insertion of a gene occurs at the repetitive region of the first inserted gene (Clare, Rayment, Ballantine, Sreekrishna, & Romanos, 1991; Fahnestock & Bedzyk, 1997), which leads to genes and subsequently to proteins with varying sizes.

An amphiphilic silk-like polymer named EL28 ($[(GA)_3GE(GA)_3GL]_{28}$; 32.4 kDa) and a nonamphiphilic silk-like polymer named EE24 ($[(GA)_3GE(GA)_3GE]_{24}$; 28.2 kDa) have been secreted using *P. pastoris*. These silk-like polymers were used to determine the usefulness of producing a pH-responsive surface active polymer in this host organism (Werten et al., 2008). The produced EE24 was purified by isoelectric precipitation, and remaining host proteins were removed by ethanol precipitation. Using this purification strategy, 3 g/L of 98% pure protein was obtained. The amphiphilic protein EL28 was purified by dissolving the protein precipitate in formic acid and a subsequent dilution with water, which resulted in a precipitation of the host proteins, whereas the target protein remained soluble for a while. This purification strategy yielded 1 g/L of a 98% pure protein

(Werten et al., 2008). The obtained 1–3 g/L of pure protein reflect high yields for *P. pastoris* (Cregg, Cereghino, Shi, & Higgins, 2000) as well as host organisms in general (Schmidt, 2004).

Further, two synthetic genes were created, based on nucleotide sequences of cDNA of *N. clavipes* (MaSp1) and *Nephila madagascariensis* (MaSp2). Both genes, which encode a 94 kDa and a 113 kDa protein, respectively, were subcloned and expressed in *P. pastoris* (Agapov et al., 2009; Bogush et al., 2009). The proteins were purified using cation exchange chromatography followed by dialysis and lyophylization. The purified proteins were processed into meshes by electrospinning, cast into thin films, and spun into fibers using wet-spinning. The fibers obtained by wet-spinning showed an elasticity of 5–15% and a tensile strength of 0.1–0.15 GPa (Bogush et al., 2009). Possible applications of the recombinant proteins include biomedical ones, such as drug delivery and tissue engineering (Bogush et al., 2009).

Fahnestock and Bedzyk (1997) created synthetic genes encoding mimicking sequences of the two dragline proteins MaSp1 and MaSp2 of *N. clavipes*. The proteins were purified from the cell lysate by acid and heat precipitation, followed by a step-wise $(NH_4)_2SO_4$ precipitation. Of the totally produced target protein (663 mg/L), 45% was obtained after purification (with 5% impurities). Immunoblotting revealed a protein ladder, probably due to internal deletions within the gene. The obtained proteins were spun into fibers under varying conditions, but the tensile strength of natural spider silk fibers could not be gained (Fahnestock et al., 2000). Using *P. pastoris* as a host organism for the recombinant spider silk protein production was included in the above mentioned patent (Fahnestock, 1994), but no further publications were made.

3.3. Plants

Introducing and expressing synthetic spider silk genes in transgenic plants is another promising approach in order to obtain large amounts of spider silk proteins, as plants are capable of producing foreign proteins on a large scale with lower costs than most other host organisms (Barr, Fahnestock, & Yang, 2004). Additionally, the production of recombinant spidroins in plants is easily scalable. By targeting recombinant proteins to cellular compartments, such as the endoplasmic reticulum (ER), vacuole, and apoplast (Conrad & Fiedler, 1998; Moloney & Holbrook, 1997), protein degradation can be prevented, and consequently higher yields can be achieved. Besides *Arabidopsis*, tobacco (*Nicotiana tabacum*) and potato (*Solanum tuberosum*) plants have been used as hosts for the production of spider silk proteins.

Disadvantages of this system are a more complicated genetic manipulation than for bacteria and longer generation intervals (Heim et al., 2009). Furthermore, only the production of recombinant spidroins with a molecular weight up to 100 kDa is reliable, as the underlying genes of larger proteins are prone to genetic rearrangements and multiple insertions of the target gene (Barr et al., 2004).

Synthetic genes that encode silk proteins based on sequences of MaSp1 of *N. clavipes* were expressed in the leaves and seeds of *Arabidopsis thaliana*, as well as somatic soybean embryos (Barr et al., 2004). A 64 kDa silk-like protein was produced successfully, whereas the production of a 127 kDa protein yielded not only the target protein but many smaller by-products as well, being caused by genetic rearrangements and multiple insertions of the target genes (Barr et al., 2004). Further, in contrast to the 64 kDa silk-like protein, the 127 kDa protein was not detectable in the somatic soybean embryos (Barr et al., 2004).

Targeting of these proteins to special cellular compartments, such as the ER and apoplasts in *Arabidopsis* leaves, increased protein yields 5-fold and 13-fold, respectively, as degradation is prevented. In *Arabidopsis* seeds, targeting the recombinant spidroins to the ER and vacuole increased the yield 7.8-fold and 5.4-fold, respectively (Yang, Barr, Fahnestock, & Liu, 2005).

Recombinant *N. clavipes* dragline proteins with a molecular weight of up to 100 kDa have been accumulated in the ER of tobacco (*N. tabacum*) and potato leaves (*S. tuberosum*), as well as in potato tubers (Menassa et al., 2004; Scheller, Guhrs, Grosse, & Conrad, 2001). The extraction of the pure recombinant protein yielded up to 3 mg protein per kg of tissue (Menassa et al., 2004). In order to simplify purification, hybrid systems of silk and another structural protein, namely elastin, have been developed. Synthetic elastin-like polypeptides, which comprise repeats of Val-Pro-Gly-Xaa-Gly (where Xaa can be any amino acid but proline), can undergo inverse temperature transition. Such synthetic polypeptide is soluble in water below 25 °C, but when the temperature is raised above 25 °C, the polypeptide aggregates, and phase separation occurs (Meyer & Chilkoti, 1999; Urry, Haynes, Zhang, Harris, & Prasad, 1988). By fusing such elastin-like domains to the target silk protein, selective precipitation can be achieved. The extraction of a spider silk-elastin fusion protein out of 1 kg of tobacco leaves yielded 80 mg of pure recombinant protein (Scheller & Conrad, 2005; Scheller, Henggeler, Viviani, & Conrad, 2004).

3.4. Insect Cells

Introducing foreign genes into higher organisms can be achieved through different procedures, such as the integration of the synthetic genes into the chromosomal DNA or by transporting the target DNA into the cell, resulting in transient expression (Maeda, 1989). Since the transformation efficiency is low with both approaches, viruses are much more promising as a transporter, as they are extraordinarily efficient in transferring their own genome into foreign cells. Using the baculovirus as a vector has been widely used in the successful expression of target genes at high levels (Maeda, 1989). For this purpose, both *Autographa californica* nuclear polyhedrosis virus (AcNPV) (Pennock, Shoemaker, & Miller, 1984) and *B. mori* NPV (BmNPV)(Maeda et al., 1985) have been established. The use of BmNPV has several advantages, since the natural host of BmNPV is the silkworm *B. mori* (Maeda, 1989). Using the insect cells for the recombinant production of spider silk proteins has several advantages, one being that insects are phylogenetically closest related to spiders compared to other commonly used expression systems (Huemmerich, Scheibel, et al., 2004). In addition, translational stops or pauses, which result in truncated proteins, were not observed using insect cells as host organisms. Other advantages include high-expression efficiencies, low feeding costs, the capability to secrete proteins simplifying the purification, as well as the ability to posttranslationally modify the proteins (Heim et al., 2009; Miao et al., 2006).

Disadvantages include more complex cloning procedures and longer generation times compared to bacteria. In addition, some spider silk proteins formed aggregates in the cytosol of the cells, which complicates the purification and subsequently lowers the protein yield (Heim et al., 2009; Huemmerich, Scheibel, et al., 2004).

Using the baculovirus in combination with the insect cell line Sf9, derived from the fall armyworm *Spodoptera frugiperda* (Huemmerich, Scheibel, et al., 2004), partial cDNAs encoding 35–105 kDa MA spidroins from *A. diadematus* (ADF3 and/or ADF4 ± carboxy-terminal domains) have been expressed (Huemmerich, Scheibel, et al., 2004; Ittah, Barak, & Gat, 2010; Ittah, Cohen, Garty, Cohn, & Gat, 2006; Ittah, Michaeli, Goldblum, & Gat, 2007). In contrast to ADF3, which was soluble in the cytosol of the insect cell, ADF4 was insoluble and self-assembled into filaments. This difference is assumed to be based on different properties of the proteins, such as hydropathicity (Huemmerich, Scheibel, et al., 2004). Contrary to other expression systems, only few degradation products or smaller protein fragments were detected, which confirms that no translational pauses or stops take place.

The baculovirus expression system using AcNPV and the insect cell line Sf9 was employed to produce a 28 kDa *Araneus ventricosus* Flagelliform protein (AvFlag), as well as a 61 kDa polyhedron-AvFlag fusion protein (Lee et al., 2007), but no information was given concerning the yield of the proteins.

The BmNPV expression system was used to produce a 37 kDa-recombinant spidroin based on the silk of *N. clavipes* in BmN cells (Miao et al., 2006). The target protein was purified from the BmN cell lysate by using Ni-NTA spin columns under denaturing conditions. Analysis of the eluted fractions showed the 37 kDa protein, as well as a 74 kDa protein, which is assumed to result from dimerization of the 37 kDa protein. The 37 kDa protein was obtained with yields of 13.3 mg/L (Miao et al., 2006). Information about chemical and structural characteristics of the produced protein has not been published so far.

3.5. Mammalian Cells

When producing recombinant spider silk proteins in different host organisms, truncated synthesis especially of high molecular weight proteins is the limiting factor. In order to overcome this limitation, mammalian cells were used (Grip et al., 2006; Lazaris et al., 2002). Additionally, mammalian cells are capable of secreting the produced proteins, which simplifies purification and enables higher protein yields.

Even though mammalian cells are able to produce larger proteins than bacteria, problems with the transcription and translation of the highly repetitive spider silk genes have been reported (Grip et al., 2006). Inefficient transcription due to the secondary structure of the mRNA, low copy number transfection of the target constructs and limitations of the cell's translational machinery were listed as possible reasons for the low expression levels. Aggregation of the produced proteins and insufficient secretion of large proteins were stated as reasons for the low protein yield (Grip et al., 2006; Lazaris et al., 2002).

Bovine mammary epithelial alveolar (MAC-T) and baby hamster kidney (BHK) cells have been tested for the expression of spider dragline silk cDNAs from two spider species, ADF3 from *A. diadematus* and MaSp1 and MaSp2 from *N. clavipes*, (molecular weights ranging from 60 kDa to 140 kDa)(Lazaris et al., 2002). Even though 110 and 140 kDa proteins were secreted to the extracellular environment, their yields in BHK cells were much lower than that of the 60 kDa protein.

Different constructs of a partial cDNA fragment of MaSp1 of *Euprosthenops* sp. were expressed in mammalian cells (COS-1) in order to provide recombinant spider silk proteins as a biomaterial for bone replacement after tumor surgery. With this approach, low expression levels were obtained caused by the above stated reasons (Grip et al., 2006).

3.6. Transgenic Animals

Transgenic animals have also been tested for the production of recombinant spider silk proteins. Advantages are possible PTMs and secretion into the milk or urine of transgenic goats and mice, which enables protein production for long time intervals, thus gaining a higher protein yield (Heim et al., 2009; Karatzas, Turner, & Karatzas, 1999). Additionally, the produced proteins can be directly spun as fibers as in the case of transgenic silkworms (Wen et al., 2010).

The process of using transgenic mammals for recombinant spider silk production has been patented (Karatzas et al., 1999). Importantly, there are a lot of disadvantages of such system that have to be considered. Firstly, the creation of transgenic mammals is very time-consuming and complex. Secondly, obtaining the spider silk proteins that were secreted into the milk of goats and mice is challenging, as the spider silk proteins have to be separated from milk caseins, and especially mice only produce a small amount of milk (Heim et al., 2009; Xu et al., 2007).

3.6.1. Transgenic Silkworms

Using a Bac-to-Bac/BmNPV Baculovirus expression system, a 70 kDa-fusion protein consisting of a recombinant spidroin (MaSp1) from *Nephila clavata*, fused to EGFP, was co-produced with silk fibroin in BmN cells as well as in the transgenic silkworm larva (Zhang et al., 2008). As 60% of the produced fusion protein was found to be insoluble, and the silkworm was not able to assemble the spider silk proteins into fibers, this approach is not feasible to obtain large amount of spider silk proteins. In order to overcome these difficulties, transgenic silkworms were created that produce a chimeric silkworm and spider silk protein (Chung, Kim, & Lee, 2012). Using this approach, composite silk fibers can be obtained, which are tougher than the parental silkworm silk fibers, and that are mechanically comparable to natural MA spider silk fibers (Teule et al., 2012).

In another approach, a vector was created that carries a partial cDNA encoding MaSp1 (83 kDa) of *N. clavata* under the sericin promoter (Ser1) (Wen et al., 2010). Transgenic silkworms grown from eggs that were injected with the altered vector spun a cocoon that contained recombinant spider silk that was located in the sericin layer of the original silkworm silk. A comparison of the mechanical properties of wild-type silkworm silk with transgenic silkworm silk and natural spider silk showed that the silk from the transgenic silkworm (18.5% elasticity, 660 MPa tensile strength) was superior to natural silkworm silk (15.3% elasticity, 564 MPa tensile strength) but did not reach the extraordinary properties of natural spider silk fibers (30% elasticity, 1300 MPa tensile strength)(Wen et al., 2010).

3.6.2. Transgenic Mice and Goats

Transgenic mice have been used to express synthetic genes based on the partial cDNA of the MaSp1/MaSp2 dragline proteins from *N. clavipes*. It was found that some mice produced a 55 kDa target protein, whereas other mice produced inhomogenous proteins with molecular weights of 31, 45, and 66 kDa. The different molecular weights are assumed to be caused by an error during protein synthesis. Yields of up to 11.7 mg/L were obtained, and the synthetic genes were stably transmitted to the mice's offspring (Xu et al., 2007).

Two recombinant 50 kDa proteins, whose encoding DNA is based on the genes of MaSp1 and MaSp2 of *N. clavipes*, were produced in the milk of transgenic goats. These proteins were purified by tangential flow filtration and chromatography, followed by alcohol precipitation. The purity of the obtained proteins was >95%. Four different w/w-ratios of the two proteins (100:0, 70:30, 30:70, 0:100 (% MaSp1 analog − % MaSp2 analog)) were solved in hexafluoroisopropanol (HFIP) to prepare individual spinning dopes. By extruding the spinning dope into a 2-propanol coagulation bath followed by post-spin drawing, recombinant fibers were produced. It was shown that the tensile strength of the artificial fibers made from all dopes (280–350 MPa) were considerably lower than that of the natural spider silk fibers (1800 MPa), whereas their elasticities were in the same range (30–40% for artificial and 26% for natural spider silk fibers) (Perez-Rigueiro et al., 2011).

4. CONCLUSION AND OUTLOOK

Natural spider silk fibers have outstanding mechanical properties with a combination of stability and extensibility resulting in a toughness accomplished by no other fibrous material. Due to its extraordinary properties, spider silk is eligible to be used in various textile, automotive, and medical engineering applications.

The biotechnological production of spider silk proteins is essential in order to investigate and to employ them for applications, because in contrast to the silkworm *B. mori*, the spider's aggressive territorial behavior and its cannibalism render their farming not feasible. The expression of partial cDNAs in different host organisms (Table 4.3) has only led to limited success (protein yields: 4–50 mg/L), as differences in the codon usages cause inefficient translations, and the highly repetitive nature of spider silk genes impedes its gene manipulation and amplification (Heim et al., 2009; Scheibel, 2004).

In order to overcome these problems, synthetic genes were produced encoding proteins differing from the natural spider silk proteins but possessing their key features. The expression of synthetic spider silk genes in different host organisms resulted in protein yields ranging from 2 mg/L (Bini et al., 2006) to 300 mg/L (Fahnestock & Bedzyk, 1997), with one exception of 3 g/L (Werten et al., 2008). Common problems are inefficient transcription due to the secondary structure of the mRNA and limitations of the cells' translational machinery, such as depletion of tRNA pools due to the highly repetitive nature of the spider silk genes. Heterologous proteins caused by truncations and gene instability are more pronounced in prokaryotes, but occur in eukaryotes as well, a problem rising with increasing molecular weights of the proteins. Secretion of the produced proteins into the extracellular environment simplifies the purification and results in higher yields, but as the underlying mechanism is more complex than that of cytosolic production, this system is more time-consuming and more prone to errors.

For many applications, using *E. coli* as a host organism for the recombinant spider silk protein production seems most promising, as the simple genetic manipulation and the short generation times enable fast adjustments. For biomedical applications such as tissue engineering and drug delivery, the engineered spider silk proteins are processed mostly into films (Slotta et al., 2006; Wohlrab et al., 2012), nonwovens (Leal-Egana et al., 2012), hydrogels (Schacht & Scheibel, 2011), and particles (Lammel, Hu, Park, Kaplan, & Scheibel, 2010). In contrast to fibers, where the molecular weight has a decisive influence on the properties concerning both the processing of the proteins and the fabricated fiber, the molecular weight contributes little to the properties of these morphologies.

Until today, recombinant spider silk proteins produced by metabolically engineered *E. coli* or transgenic goats were spun into fibers with mechanical properties, that were similar but still lower than those of natural spider silk fibers. Therefore, gaining artificial silk fibers with mechanical properties as found as in nature still remains the holy grail of silk research.

ACKNOWLEDGMENTS

We greatly acknowledge financial funding of DFG SCHE 603/4-4.

REFERENCES

Agapov, II, Pustovalova, O. L., Moisenovich, M. M., Bogush,V. G., Sokolova, O. S., Sevastyanov, V. I., et al. (2009).Three-dimensional scaffold made from recombinant spider silk protein for tissue engineering. *Doklady Biochemistry and Biophysics, 426*, 127–130.

Andersen, S. O. (1970).Amino acid composition of spider silks. *Comparative Biochemistry and Physiology, 35*, 705–711.

Ap Rhisiart, A., & Vollrath, F. (1994). Design features of the orb web of the spider, *Araneus diadematus. Behavioral Ecology, 5*, 280–287.

Arcidiacono, S., Mello, C. M., Butler, M., Welsh, E., Soares, J. W., Allen, A., et al. (2002). Aqueous processing and fiber spinning of recombinant spider silks. *Macromolecules, 35*, 1262–1266.

Arcidiacono, S., Mello, C., Kaplan, D., Cheley, S., & Bayley, H. (1998). Purification and characterization of recombinant spider silk expressed in *Escherichia coli. Applied Microbiology and Biotechnology, 49*, 31–38.

Asakura, T., Kuzuhara, A., Tabeta, R., & Saito, H. (1985). Conformation characterization of *Bombyx mori* silk fibroin in the solid state by high-frequency C-13 cross polarization-magic angle spinning NMR, X-ray diffraction, and infrared spectroscopy. *Macromolecules, 18*, 1841–1845.

Askarieh, G., Hedhammar, M., Nordling, K., Saenz, A., Casals, C., Rising, A., et al. (2010). Self-assembly of spider silk proteins is controlled by a pH-sensitive relay. *Nature, 465*, 236–238.

Augsten, K., Muhlig, P., & Herrmann, C. (2000). Glycoproteins and skin-core structure in *Nephila clavipes* spider silk observed by light and electron microscopy. *Scanning, 22*, 12–15.

Augsten, K., Weisshart, K., Sponner, A., & Unger, E. (1999). Glycoproteins and skin-core structure in *Nephila clavipes* spider silk observed by light- and electron microscopy. *Scanning, 21*, 77.

Ayoub, N. A., Garb, J. E.,Tinghitella, R. M., Collin, M. A., & Hayashi, C.Y. (2007). Blueprint for a high-performance biomaterial: full-length spider dragline silk genes. *PLoS ONE, 2*, e514.

Barghout, J.Y.J., Thiel, B. L., & Viney, C. (1999). Spider (*Araneus diadematus*) cocoon silk: a case of non-periodic lattice crystals with a twist? *International Journal of Biological Macromolecules, 24*, 211–217.

Barr, L. A., Fahnestock, S. R., & Yang, J. J. (2004). Production and purification of recombinant DP1B silk-like protein in plants. *Molecular Breeding, 13*, 345–356.

Becker, N., Oroudjev, E., Mutz, S., Cleveland, J. P., Hansma, P. K., Hayashi, C.Y., et al. (2003). Molecular nanosprings in spider capture-silk threads. *Nature Materials, 2*, 278–283.

Bini, E., Foo, C.W., Huang, J., Karageorgiou,V., Kitchel, B., & Kaplan, D. L. (2006). RGD-functionalized bioengineered spider dragline silk biomaterial. *Biomacromolecules, 7*, 3139–3145.

Bini, E., Knight, D. P., & Kaplan, D. L. (2004). Mapping domain structures in silks from insects and spiders related to protein assembly. *Journal of Molecular Biology, 335*, 27–40.

Bittencourt, D., Souto, B. M.,Verza, N. C.,Vineck, F., Dittmar, K., Silva, P. I., et al. (2007). Spidroins from the Brazilian spider *Nephilengys cruentata* (Araneae: nephilidae). *Comparative Biochemistry and Physiology. Part B, Biochemistry & Molecular Biology, 147*, 597–606.

Blackledge, T. A., & Hayashi, C.Y. (2006). Silken toolkits: biomechanics of silk fibers spun by the orb web spider *Argiope argentata* (Fabricius 1775). *Journal of Experimental Biology, 209*, 2452–2461.

Blackledge, T. A., Summers, A. P., & Hayashi, C.Y. (2005). Gumfooted lines in black widow cobwebs and the mechanical properties of spider capture silk. *Zoology, 108*, 41–46.

Bogush, V. G., Sokolova, O. S., Davydova, L. I., Klinov, D. V., Sidoruk, K. V., Esipova, N. G., et al. (2009). A novel model system for design of biomaterials based on recombinant analogs of spider silk proteins. *Journal of Neuroimmune Pharmacology, 4,* 17–27.

Bon, M. (1710). A discourse upon the usefulness of the silk of spiders. *Philosophical Transactions, 27,* 2–16.

Brooks, A. E., Nelson, S. R., Jones, J. A., Koenig, C., Hinman, M., Stricker, S., et al. (2008). Distinct contributions of model MaSp1 and MaSp2 like peptides to the mechanical properties of synthetic major ampullate silk fibers as revealed in silico. *Nanotechnology, Science and Applications, 1,* 9–16.

Brooks, A. E., Steinkraus, H. B., Nelson, S. R., & Lewis, R. V. (2005). An investigation of the divergence of major ampullate silk fibers from *Nephila clavipes* and *Argiope aurantia*. *Biomacromolecules, 6,* 3095–3099.

Brooks, A. E., Stricker, S. M., Joshi, S. B., Kamerzell, T. J., Middaugh, C. R., & Lewis, R. V. (2008). Properties of synthetic spider silk fibers based on *Argiope aurantia* MaSp2. *Biomacromolecules, 9,* 1506–1510.

Candelas, G. C., & Cintron, J. (1981). A spider fibroin and its synthesis. *Journal of Experimental Zoology, 216,* 1–6.

Cappello, J., & Crissman, J. W. (1990). The design and production of bioactive protein polymers for biomedical applications. *Abstracts of Papers of the American Chemical Society, 199,* 66.

Cappello, J., Crissman, J., Dorman, M., Mikolajczak, M., Textor, G., Marquet, M., et al. (1990). Genetic engineering of structural protein polymers. *Biotechnology Progress, 6,* 198–202.

Cereghino, G. P., Cereghino, J. L., Ilgen, C., & Cregg, J. M. (2002). Production of recombinant proteins in fermenter cultures of the yeast *Pichia pastoris*. *Current Opinion in Biotechnology, 13,* 329–332.

Challis, R. J., Goodacre, S. L., & Hewitt, G. M. (2006). Evolution of spider silks: conservation and diversification of the C-terminus. *Insect Molecular Biology, 15,* 45–56.

Chung, H., Kim, T. Y., & Lee, S. Y. (2012). Recent advances in production of recombinant spider silk proteins. *Current Opinion in Biotechnology, 23,* 1–8.

Clare, J. J., Rayment, F. B., Ballantine, S. P., Sreekrishna, K., & Romanos, M. A. (1991). High-level expression of tetanus toxin fragment C in *Pichia pastoris* strains containing multiple tandem integrations of the gene. *Biotechnology, 9,* 455–460.

Colgin, M. A., & Lewis, R. V. (1998). Spider minor ampullate silk proteins contain new repetitive sequences and highly conserved non-silk-like "spacer regions". *Protein Science, 7,* 667–672.

Conrad, U., & Fiedler, U. (1998). Compartment-specific accumulation of recombinant immunoglobulins in plant cells: an essential tool for antibody production and immunomodulation of physiological functions and pathogen activity. *Plant Molecular Biology, 38,* 101–109.

Craig, C. L., Riekel, C., Herberstein, M. E., Weber, R. S., Kaplan, D., & Pierce, N. E. (2000). Evidence for diet effects on the composition of silk proteins produced by spiders. *Molecular Biology and Evolution, 17,* 1904–1913.

Cregg, J. M. (Ed.), (2007). *Pichia protocols*. Totowa, New Jersey: Humana Press Inc.

Cregg, J. M., Cereghino, J. L., Shi, J. Y., & Higgins, D. R. (2000). Recombinant protein expression in *Pichia pastoris*. *Molecular Biotechnology, 16,* 23–52.

Dicko, C., Knight, D., Kenney, J. M., & Vollrath, F. (2004). Secondary structures and conformational changes in flagelliform, cylindrical, major, and minor ampullate silk proteins. Temperature and concentration effects. *Biomacromolecules, 5,* 2105–2115.

Eisoldt, L., Hardy, J. G., Heim, M., & Scheibel, T. R. (2010). The role of salt and shear on the storage and assembly of spider silk proteins. *Journal of Structural Biology, 170,* 413–419.

Eisoldt, L., Smith, A. & Scheibel, T. (2011). Decoding the secrets of spider silk. *Materials Today, 14,* 80–86.

Eisoldt, L., Thamm, C., & Scheibel, T. (2011). The role of terminal domains during storage and assembly of spider silk proteins. *Biopolymers, 97,* 355–361.

Exler, J. H., Hummerich, D., & Scheibel, T. (2007). The amphiphilic properties of spider silks are important for spinning. *Angewandte Chemie International Edition*, *46*, 3559–3562.

Fahnestock, S. (1994). In W. E. I. DuPont (Ed.), *Novel, recombinantly produced spider silk analogs*. USA, int. patent number: WO 94/29450.

Fahnestock, S. R., & Bedzyk, L. A. (1997). Production of synthetic spider dragline silk protein in *Pichia pastoris*. *Applied Microbiology and Biotechnology*, *47*, 33–39.

Fahnestock, S. R., & Irwin, S. L. (1997). Synthetic spider dragline silk proteins and their production in *Escherichia coli*. *Applied Microbiology and Biotechnology*, *47*, 23–32.

Fahnestock, S. R., Yao, Z., & Bedzyk, L. A. (2000). Microbial production of spider silk proteins. *Journal of Biotechnology*, *74*, 105–119.

Fox, L. R. (1975). Cannibalism in natural populations. *Annual Review of Ecology and Systematics*, *6*, 87–106.

Fukushima, Y. (1998). Genetically engineered syntheses of tandem repetitive polypeptides consisting of glycine-rich sequence of spider dragline silk. *Biopolymers*, *45*, 269–279.

Galan, J. E., & Collmer, A. (1999). Type III secretion machines: bacterial devices for protein delivery into host cells. *Science*, *284*, 1322–1328.

Garb, J. E., & Hayashi, C. Y. (2005). Modular evolution of egg case silk genes across orb-weaving spider superfamilies. *Proceedings of the National Academy of Sciences of the United States of America*, *102*, 11379–11384.

Gellissen, G., Melber, K., Janowicz, Z. A., Dahlems, U. M., Weydemann, U., Piontek, M., et al. (1992). Heterologous protein production in yeast. *Antonie Van Leeuwenhoek International Journal of General and Molecular Microbiology*, *62*, 79–93.

Gerritsen, V. B. (2002). The tiptoe of an airbus. *Protein Spotlight, Swiss Prot*, *24*, 1–2.

Geurts, P., Zhao, L., Hsia, Y., Gnesa, E., Tang, S., Jeffery, F., et al. (2010). Synthetic spider silk fibers spun from pyriform spidroin 2, a glue silk protein discovered in orb-weaving spider attachment discs. *Biomacromolecules*, *11*, 3495–3503.

Gnesa, E., Hsia, Y., Yarger, J. L., Weber, W., Lin-Cereghino, J., Lin-Cereghino, G., et al. (2012). Conserved C-terminal domain of spider tubuliform spidroin 1 contributes to extensibility in synthetic fibers. *Biomacromolecules*, *13*, 304–312.

Gosline, J. M., Denny, M. W., & DeMont, M. E. (1984). Spider silk as rubber. *Nature*, *309*, 551–552.

Gosline, J. M., Guerette, P. A., Ortlepp, C. S., & Savage, K. N. (1999). The mechanical design of spider silks: from fibroin sequence to mechanical function. *Journal of Experimental Biology*, *202*, 3295–3303.

Gosline, J., Lillie, M., Carrington, E., Guerette, P., Ortlepp, C., & Savage, K. (2002). Elastic proteins: biological roles and mechanical properties. *Philosophical Transactions of the Royal Society of London. Series B, Biological Sciences*, *357*, 121–132.

Gosline, J. M., Pollak, C. C., Guerette, P. A., Cheng, A., DeMont, M. E., & Denny, M. W. (1993). Elastomeric network models for the frame and viscid silks from the orb web of the spider *Araneus diadematus*. *Silk Polymers*, *544*, 328–341.

Grip, S., Rising, A., Nimmervoll, H., Storckenfeldt, E., Mcqueen-Mason, S. J., Pouchkina-Stantcheva, N., et al. (2006). Transient expression of a major ampullate spidroin 1 gene fragment from *Euprosthenops* sp. in mammalian cells. *Cancer Genomics & Proteomics*, *3*, 83–87.

Guerette, P. A., Ginzinger, D. G., Weber, B. H., & Gosline, J. M. (1996). Silk properties determined by gland-specific expression of a spider fibroin gene family. *Science*, *272*, 112–115.

Hagn, F., Eisoldt, L., Hardy, J. G., Vendrely, C., Coles, M., Scheibel, T., et al. (2010). A conserved spider silk domain acts as a molecular switch that controls fibre assembly. *Nature*, *465*, 239–242.

Hagn, F., Thamm, C., Scheibel, T., & Kessler, H. (2010). pH-dependent dimerization and salt-dependent stabilization of the N-terminal domain of spider dragline silk – implications for fiber formation. *Angewandte Chemie International Edition*, *49*, 1–5.

Hajer, J., & Rehakova, D. (2003). Spinning activity of the spider *Trogloneta granulum* (Araneae, Mysmenidae): web, cocoon, cocoon handling behaviour, draglines and attachment discs. *Zoology, 106*, 223–231.

Hardy, J. G., & Scheibel, T. R. (2009). Silk-inspired polymers and proteins. *Biochemical Society Transactions, 37*, 677–681.

Harvey, B. R., Georgiou, G., Hayhurst, A., Jeong, K. J., Iverson, B. L., & Rogers, G. K. (2004). Anchored periplasmic expression, a versatile technology for the isolation of high-affinity antibodies from *Escherichia coli*-expressed libraries. *Proceedings of the National Academy of Sciences of the United States of America, 101*, 9193–9198.

Hawthorn, A. C., & Opell, B. D. (2002). Evolution of adhesive mechanisms in cribellar spider prey capture thread: evidence for van der Waals and hygroscopic forces. *Biological Journal of the Linnean Society, 77*, 1–8.

Hawthorn, A. C., & Opell, B. D. (2003). Van der Waals and hygroscopic forces of adhesion generated by spider capture threads. *Journal of Experimental Biology, 206*, 3905–3911.

Hayashi, C. Y., Blackledge, T. A., & Lewis, R. V. (2004). Molecular and mechanical characterization of aciniform silk: uniformity of iterated sequence modules in a novel member of the spider silk fibroin gene family. *Molecular Biology and Evolution, 21*, 1950–1959.

Hayashi, C. Y., & Lewis, R. V. (1998). Evidence from flagelliform silk cDNA for the structural basis of elasticity and modular nature of spider silks. *Journal of Molecular Biology, 275*, 773–784.

Hayashi, C. Y., Shipley, N. H., & Lewis, R. V. (1999). Hypotheses that correlate the sequence, structure, and mechanical properties of spider silk proteins. *International Journal of Biological Macromolecules, 24*, 271–275.

Hedhammar, M., Rising, A., Grip, S., Martinez, A. S., Nordling, K., Casals, C., et al. (2008). Structural properties of recombinant nonrepetitive and repetitive parts of major ampullate spidroin 1 from *Euprosthenops australis*: implications for fiber formation. *Biochemistry, 47*, 3407–3417.

Heim, M., Ackerschott, C. B., & Scheibel, T. (2010). Characterization of recombinantly produced spider flagelliform silk domains. *Journal of Structural Biology, 170*, 420–425.

Heim, M., Keerl, D., & Scheibel, T. (2009). Spider silk: from soluble protein to extraordinary fiber. *Angewandte Chemie International Edition, 48*, 3584–3596.

Hijirida, D. H., Do, K. G., Michal, C., Wong, S., Zax, D., & Jelinski, L. W. (1996). 13C NMR of *Nephila clavipes* major ampullate silk gland. *Biophysical Journal, 71*, 3442–3447.

Hinman, M. B., & Lewis, R. V. (1992). Isolation of a clone encoding a second dragline silk fibroin. *Nephila clavipes* dragline silk is a two-protein fiber. *Journal of Biological Chemistry, 267*, 19320–19324.

Holland, C., Terry, A. E., Porter, D., & Vollrath, F. (2007). Natural and unnatural silks. *Polymer, 48*, 3388–3392.

Huang, W., Lin, Z., Sin, Y. M., Li, D., Gong, Z., & Yang, D. (2006). Characterization and expression of a cDNA encoding a tubuliform silk protein of the golden web spider *Nephila antipodiana*. *Biochimie, 88*, 849–858.

Huang, J., Wong, C., George, A., & Kaplan, D. L. (2007). The effect of genetically engineered spider silk-dentin matrix protein 1 chimeric protein on hydroxyapatite nucleation. *Biomaterials, 28*, 2358–2367.

Huemmerich, D., Helsen, C. W., Quedzuweit, S., Oschmann, J., Rudolph, R., & Scheibel, T. (2004). Primary structure elements of spider dragline silks and their contribution to protein solubility. *Biochemistry, 43*, 13604–13612.

Huemmerich, D., Scheibel, T., Vollrath, F., Cohen, S., Gat, U., & Ittah, S. (2004). Novel assembly properties of recombinant spider dragline silk proteins. *Current Biology, 14*, 2070–2074.

Huemmerich, D., Slotta, U., & Scheibel, T. (2006). Processing and modification of films made from recombinant spider silk proteins. *Applied Physics. A, -Materials Science & Processing, 82*, 219–222.

Hu, X. Y., Kohler, K., Falick, A. M., Moore, A. M.F., Jones, P. R., Sparkman, O. D., et al. (2005). Egg case protein-1-A new class of silk proteins with fibroin-like properties from the spider *Latrodectus hesperus*. *Journal of Biological Chemistry*, *280*, 21220–21230.

Hu, X., Kohler, K., Falick, A. M., Moore, A. M., Jones, P. R., & Vierra, C. (2006). Spider egg case core fibers: trimeric complexes assembled from TuSp1, ECP-1, and ECP-2. *Biochemistry*, *45*, 3506–3516.

Hu, X. Y., Lawrence, B., Kohler, K., Falick, A. M., Moore, A. M.F., McMullen, E., et al. (2005). Araneoid egg case silk: a fibroin with novel ensemble repeat units from the black widow spider, *Latrodectus hesperus*. *Biochemistry*, *44*, 10020–10027.

Hurley, J. H., Mason, D. A., & Matthews, B. W. (1992). Flexible-geometry conformational energy maps for the amino acid residue preceding a proline. *Biopolymers*, *32*, 1443–1446.

Hu, X., Vasanthavada, K., Kohler, K., McNary, S., Moore, A. M., & Vierra, C. A. (2006). Molecular mechanisms of spider silk. *Cellular and Molecular Life Sciences*, *63*, 1986–1999.

Hu, X. Y., Yuan, J., Wang, X. D., Vasanthavada, K., Falick, A. M., Jones, P. R., et al. (2007). Analysis of aqueous glue coating proteins on the silk fibers of the cob weaver, *Latrodectus hesperus*. *Biochemistry*, *46*, 3294–3303.

Iridag, Y., & Kazanci, M. (2006). Preparation and characterization of *Bombyx mori* silk fibroin and wool keratin. *Journal of Applied Polymer Science*, *100*, 4260–4264.

Ito, H., Muraoka, Y., Yamazaki, T., Imamura, T., Mori, H., Ichida, M., et al. (1995). Structure and chemical composition of silk proteins in relation to silkworm diet. *Textile Research Journal*, *65*, 755–759.

Ittah, S., Barak, N., & Gat, U. (2010). A proposed model for dragline spider silk self-assembly: insights from the effect of the repetitive domain size on fiber properties. *Biopolymers*, *93*, 458–468.

Ittah, S., Cohen, S., Garty, S., Cohn, D., & Gat, U. (2006). An essential role for the C-terminal domain of a dragline spider silk protein in directing fiber formation. *Biomacromolecules*, *7*, 1790–1795.

Ittah, S., Michaeli, A., Goldblum, A., & Gat, U. (2007). A model for the structure of the C-terminal domain of dragline spider silk and the role of its conserved cysteine. *Biomacromolecules*, *8*, 2768–2773.

Jackson, C., & O'Brien, J. P. (1995). Molecular weight distribution of *Nephila clavipes* dragline silk. *Macromolecules*, *28*, 5975–5977.

Jelinski, L. W., Blye, A., Liivak, O., Michal, C., LaVerde, G., Seidel, A., et al. (1999). Orientation, structure, wet-spinning, and molecular basis for supercontraction of spider dragline silk. *International Journal of Biological Macromolecules*, *24*, 197–201.

Karatzas, C. N., Turner, J. D., & Karatzas, A.-L. (1999). Production of biofilaments in transgenic animals. *Canada*, int. patent number: WO 99/47661.

Knight, D., & Vollrath, F. (1999). Hexagonal columnar liquid crystal in the cells secreting spider silk. *Tissue & Cell*, *31*, 617–620.

Knight, D. P., & Vollrath, F. (2001). Changes in element composition along the spinning duct in a *Nephila* spider. *Naturwissenschaften*, *88*, 179–182.

Kohler, K., Thayer, W., Le, T., Sembhi, A., Vasanthavada, K., Moore, A. M. F., et al. (2005). Characterization of a novel class II bHLH transcription factor from the black widow spider, *Latrodectus hesperus*, with silk-gland restricted patterns of expression. *DNA and Cell Biology*, *24*, 371–380.

Kovoor, J., & Zylberberg, L. (1980). Fine structural aspects of silk secretion in a spider (*Araneus diadematus*). I. Elaboration in the pyriform glands. *Tissue Cell*, *12*, 547–556.

Kukuruzinska, M. A., & Lennon, K. (1998). Protein N-glycosylation: molecular genetics and functional significance. *Critical Reviews in Oral Biology and Medicine*, *9*, 415–448.

La Mattina, C., Reza, R., Hu, X., Falick, A. M., Vasanthavada, K., McNary, S., et al. (2008). Spider minor ampullate silk proteins are constituents of prey wrapping silk in the cob weaver *Latrodectus hesperus*. *Biochemistry*, *47*, 4692–4700.

Lammel, A. S., Hu, X., Park, S. H., Kaplan, D. L., & Scheibel, T. R. (2010). Controlling silk fibroin particle features for drug delivery. *Biomaterials, 31*, 4583–4591.

Lazaris, A., Arcidiacono, S., Huang, Y., Zhou, J. F., Duguay, F., Chretien, N., et al. (2002). Spider silk fibers spun from soluble recombinant silk produced in mammalian cells. *Science, 295*, 472–476.

Leal-Egana, A., Lang, G., Mauerer, C., Wickinghoff, J., Weber, M., Geimer, S., et al. (2012). Interactions of fibroblasts with different morphologies made of an engineered spider silk protein. *Advanced Engineering Materials, 14*, B67–B75.

Lee, K. S., Kim, B. Y., Je, Y. H., Woo, S. D., Sohn, H. D., & Jin, B. R. (2007). Molecular cloning and expression of the C-terminus of spider flagelliform silk protein from *Araneus ventricosus*. *Journal of Biosciences, 32*, 705–712.

Lee, P. A., Tullman-Ercek, D., & Georgiou, G. (2006). The bacterial twin-arginine translocation pathway. *Annual Review of Microbiology, 60*, 373–395.

Lewis, R. V., Hinman, M., Kothakota, S., & Fournier, M. J. (1996). Expression and purification of a spider silk protein: a new strategy for producing repetitive proteins. *Protein Expression and Purification, 7*, 400–406.

Liebmann, B., Huemmerich, D., Scheibel, T., & Fehr, M. (2008). Formulation of poorly water-soluble substances using self-assembling spider silk protein. *Colloids and Surfaces. A, Physicochemical and Engineering Aspects, 331*, 126–132.

Liivak, O., Blye, A., Shah, N., & Jelinski, L. W. (1998). A microfabricated wet-spinning apparatus to spin fibers of silk proteins. Structure–property correlations. *Macromolecules, 31*, 2947–2951.

Lin, Z., Huang, W., Zhang, J., Fan, J. S., & Yang, D. (2009). Solution structure of eggcase silk protein and its implications for silk fiber formation. *Proceedings of the National Academy of Sciences of the United States of America, 106*, 8906–8911.

Liu, Y., Shao, Z., & Vollrath, F. (2005). Relationships between supercontraction and mechanical properties of spider silk. *Nature Materials, 4*, 901–905.

Liu, Y., Shao, Z., & Vollrath, F. (2008). Elasticity of spider silks. *Biomacromolecules, 9*, 1782–1786.

Liu, Y., Sponner, A., Porter, D., & Vollrath, F. (2008). Proline and processing of spider silks. *Biomacromolecules, 9*, 116–121.

Lombardi, S. J., & Kaplan, D. L. (1990). The amino-acid-composition of major ampullate gland silk (dragline) of *Nephila clavipes* (Araneae, Tetragnathidae). *Journal of Arachnology, 18*, 297–306.

Macauley-Patrick, S., Fazenda, M. L., McNeil, B., & Harvey, L. M. (2005). Heterologous protein production using the *Pichia pastoris* expression system. *Yeast, 22*, 249–270.

Madsen, B., Shao, Z. Z., & Vollrath, F. (1999). Variability in the mechanical properties of spider silks on three levels: interspecific, intraspecific and intraindividual. *International Journal of Biological Macromolecules, 24*, 301–306.

Maeda, S. (1989). Expression of foreign genes in insects using baculovirus vectors. *Annual Review of Entomology, 34*, 351–372.

Maeda, S., Kawai, T., Obinata, M., Fujiwara, H., Horiuchi, T., Saeki, Y., et al. (1985). Production of human alpha-interferon in silkworm using a baculovirus vector. *Nature, 315*, 592–594.

Marlovits, T. C., Kubori, T., Lara-Tejero, M., Thomas, D., Unger, V. M., & Galan, J. E. (2006). Assembly of the inner rod determines needle length in the type III secretion injectisome. *Nature, 441*, 637–640.

Marsano, E., Corsini, P., Arosio, C., Boschi, A., Mormino, M., & Freddi, G. (2005). Wet spinning of *Bombyx mori* silk fibroin dissolved in N-methyl morpholine N-oxide and properties of regenerated fibres. *International Journal of Biological Macromolecules, 37*, 179–188.

Matsumoto, K., Uejima, H., Iwasaki, T., Sano, Y., & Sumino, H. (1996). Studies on regenerated protein fibers. 3. Production of regenerated silk fibroin fiber by the self-dialyzing wet spinning method. *Journal of Applied Polymer Science, 60*, 503–511.

Mello, C. M., Senecal, B., Yeung, B., Vouros, P., & Kaplan, D. L. (1994). Initial characterization of *Nephila clavipes* dragline protein. In D. L. Kaplan, W. W. Adams, B. Farmer & C. Viney (Eds.), *Silk polymers, materials science and biotechnology*. Washington, DC: American Chemical Society, *544*, 67–79.

Mello, C. M., Soares, J. W., Arcidiacono, S., & Butlers, M. M. (2004). Acid extraction and purification of recombinant spider silk proteins. *Biomacromolecules*, *5*, 1849–1852.

Menassa, R., Hong, Z., Karatzas, C. N., Lazaris, A., Richman, A., & Brandle, J. (2004). Spider dragline silk proteins in transgenic tobacco leaves: accumulation and field production. *Plant Biotechnology Journal*, *2*, 431–438.

Meyer, D. E., & Chilkoti, A. (1999). Purification of recombinant proteins by fusion with thermally-responsive polypeptides. *Nature Biotechnology*, *17*, 1112–1115.

Miao, Y. G., Zhang, Y. S., Nakagaki, K., Zhao, T. F., Zhao, A. C., Meng, Y., et al. (2006). Expression of spider flagelliform silk protein in *Bombyx mori* cell line by a novel Bac-to-Bac/BmNPV baculovirus expression system. *Applied Microbiology and Biotechnology*, *71*, 192–199.

Michal, C. A., Simmons, A. H., Chew, B. G., Zax, D. B., & Jelinski, L. W. (1996). Presence of phosphorus in *Nephila clavipes* dragline silk. *Biophysical Journal*, *70*, 489–493.

Moloney, M. M., & Holbrook, L. A. (1997). Subcellular targeting and purification of recombinant proteins in plant production systems. *Biotechnology & Genetic Engineering Reviews*, *14*, 321–336.

O'Brien, J. P., Fahnestock, S. R., Termonia, Y., & Gardner, K. C.H. (1998). Nylons from nature: synthetic analogs to spider silk. *Advanced Materials*, *10*, 1185–1195.

Ohgo, K., Kawase, T., Ashida, J., & Asakura, T. (2006). Solid-state NMR analysis of a peptide (Gly-Pro-Gly-Gly-Ala)(6)-Gly derived from a flagelliform silk sequence of *Nephila clavipes*. *Biomacromolecules*, *7*, 1210–1214.

Patel, J., Zhu, H., Menassa, R., Gyenis, L., Richman, A., & Brandle, J. (2007). Elastin-like polypeptide fusions enhance the accumulation of recombinant proteins in tobacco leaves. *Transgenic Research*, *16*, 239–249.

Pennock, G. D., Shoemaker, C., & Miller, L. K. (1984). Strong and regulated expression of *Escherichia coli* beta-galactosidase in insect cells with a baculovirus vector. *Molecular Cell Biology*, *4*, 399–406.

Perez-Rigueiro, J., Elices, M., Guinea, G.V., Plaza, G. R., Karatzas, C., Riekel, C., et al. (2011). Bioinspired fibers follow the track of natural spider silk. *Macromolecules*, *44*, 1166–1176.

Perez-Rigueiro, J., Elices, M., Plaza, G. R., & Guinea, G. V. (2007). Similarities and differences in the supramolecular organization of silkworm and spider silk. *Macromolecules*, *40*, 5360–5365.

Prince, J. T., McGrath, K. P., DiGirolamo, C. M., & Kaplan, D. L. (1995). Construction, cloning, and expression of synthetic genes encoding spider dragline silk. *Biochemistry*, *34*, 10879–10885.

Putthanarat, S., Stribeck, N., Fossey, S. A., Eby, R. K., & Adams, W. W. (2000). Investigation of the nanofibrils of silk fibers. *Polymer*, *41*, 7735–7747.

Rauscher, S., Baud, S., Miao, M., Keeley, F.W., & Pomes, R. (2006). Proline and glycine control protein self-organization into elastomeric or amyloid fibrils. *Structure*, *14*, 1667–1676.

Riekel, C., & Vollrath, F. (2001). Spider silk fibre extrusion: combined wide- and small-angle X-ray microdiffraction experiments. *International Journal of Biological Macromolecules*, *29*, 203–210.

Rising, A., Hjalm, G., Engstrom, W., & Johansson, J. (2006). N-terminal nonrepetitive domain common to dragline, flagelliform, and cylindriform spider silk proteins. *Biomacromolecules*, *7*, 3120–3124.

Rising, A., Widhe, M., Johansson, J., & Hedhammar, M. (2011). Spider silk proteins: recent advances in recombinant production, structure–function relationships and biomedical applications. *Cellular and Molecular Life Sciences*, *68*, 169–184.

Römer, L., & Scheibel, T. (2007). Basis for new material – spider silk protein. *Chemie in Unserer Zeit*, *41*, 306–314.

Rosenberg, A. H., Goldman, E., Dunn, J. J., Studier, F. W., & Zubay, G. (1993). Effects of consecutive AGG codons on translation in *Escherichia coli*, demonstrated with a versatile codon test system. *Journal of Bacteriology, 175*, 716–722.

Schacht, K., & Scheibel, T. (2011). Controlled hydrogel formation of a recombinant spider silk protein. *Biomacromolecules, 12*, 2488–2495.

Scheibel, T. (2004). Spider silks: recombinant synthesis, assembly, spinning, and engineering of synthetic proteins. *Microbial Cell Factories, 3*.

Scheller, J., & Conrad, U. (2005). Plant-based material, protein and biodegradable plastic. *Current Opinion in Plant Biology, 8*, 188–196.

Scheller, J., Guhrs, K. H., Grosse, F., & Conrad, U. (2001). Production of spider silk proteins in tobacco and potato. *Nature Biotechnology, 19*, 573–577.

Scheller, J., Henggeler, D., Viviani, A., & Conrad, U. (2004). Purification of spider silk-elastin from transgenic plants and application for human chondrocyte proliferation. *Transgenic Research, 13*, 51–57.

Schmidt, F. R. (2004). Recombinant expression systems in the pharmaceutical industry. *Applied Microbiology and Biotechnology, 65*, 363–372.

Schmidt, M., Romer, L., Strehle, M., & Scheibel, T. (2007). Conquering isoleucine auxotrophy of *Escherichia coli* BLR(DE3) to recombinantly produce spider silk proteins in minimal media. *Biotechnology Letters, 29*, 1741–1744.

Schulz, S. (2001). Composition of the silk lipids of the spider *Nephila clavipes*. *Lipids, 36*, 637–647.

Shao, Z. Z., Vollrath, F., Yang, Y., & Thogersen, H. C. (2003). Structure and behavior of regenerated spider silk. *Macromolecules, 36*, 1157–1161.

Simmons, A. H., Michal, C. A., & Jelinski, L. W. (1996). Molecular orientation and two-component nature of the crystalline fraction of spider dragline silk. *Science, 271*, 84–87.

Slotta, U. K., Rammensee, S., Gorb, S., & Scheibel, T. (2008). An engineered spider silk protein forms microspheres. *Angewandte Chemie International Edition, 47*, 4592–4594.

Slotta, U., Tammer, M., Kremer, F., Koelsch, P., & Scheibel, T. (2006). Structural analysis of spider silk films. *Supramolecular Chemistry, 18*, 465–471.

Sorensen, H. P., & Mortensen, K. K. (2005). Advanced genetic strategies for recombinant protein expression in *Escherichia coli*. *Journal of Biotechnology, 115*, 113–128.

Sponner, A., Unger, E., Grosse, F., & Weisshart, K. (2005). Differential polymerization of the two main protein components of dragline silk during fibre spinning. *Nature Materials, 4*, 772–775.

Sponner, A., Vater, W., Monajembashi, S., Unger, E., Grosse, F., & Weisshart, K. (2007). Composition and hierarchical organisation of a spider silk. *PLoS One, 2*, e998.

Sponner, A., Vater, W., Rommerskirch, W., Vollrath, F., Unger, E., Grosse, F., et al. (2005). The conserved C-termini contribute to the properties of spider silk fibroins. *Biochemical and Biophysical Research Communications, 338*, 897–902.

Stark, M., Grip, S., Rising, A., Hedhammar, M., Engstrom, W., Hjalm, G., et al. (2007). Macroscopic fibers self-assembled from recombinant miniature spider silk proteins. *Biomacromolecules, 8*, 1695–1701.

Stauffer, S. L., Coguill, S. L., & Lewis, R. V. (1994). Comparison of physical properties of 3 silks from *Nephila clavipes* and *Araneus gemmoides*. *Journal of Arachnology, 22*, 5–11.

Stephens, J. S., Fahnestock, S. R., Farmer, R. S., Kiick, K. L., Chase, D. B., & Rabolt, J. F. (2005). Effects of electrospinning and solution casting protocols on the secondary structure of a genetically engineered dragline spider silk analogue investigated via fourier transform Raman spectroscopy. *Biomacromolecules, 6*, 1405–1413.

Szela, S., Avtges, P., Valluzzi, R., Winkler, S., Wilson, D., Kirschner, D., et al. (2000). Reduction–oxidation control of beta-sheet assembly in genetically engineered silk. *Biomacromolecules, 1*, 534–542.

Terpe, K. (2006). Overview of bacterial expression systems for heterologous protein production: from molecular and biochemical fundamentals to commercial systems. *Applied Microbiology and Biotechnology, 72*, 211–222.

Teule, F., Aube, C., Ellison, M., & Abbott, A. (2003). Biomimetic manufacturing of customised novel fibre proteins for specialised applications. *AUTEX Research Journal, 3*, 160–165.

Teule, F., Cooper, A. R., Furin, W. A., Bittencourt, D., Rech, E. L., Brooks, A., et al. (2009). A protocol for the production of recombinant spider silk-like proteins for artificial fiber spinning. *Nature Protocols, 4*, 341–355.

Teule, F., Furin, W. A., Cooper, A. R., Duncan, J. R., & Lewis, R. V. (2007). Modifications of spider silk sequences in an attempt to control the mechanical properties of the synthetic fibers. *Journal of Materials Science, 42*, 8974–8985.

Teule, F., Miao, Y. G., Sohn, B. H., Kim, Y. S., Hull, J. J., Fraser, M. J., Jr., et al. (2012). Silkworms transformed with chimeric silkworm/spider silk genes spin composite silk fibers with improved mechanical properties. *Proceedings of the National Academy of Sciences of the United States of America, 109*, 923–928.

Thiel, B. L., Guess, K. B., & Viney, C. (1997). Non-periodic lattice crystals in the hierarchical microstructure of spider (major ampullate) silk. *Biopolymers, 41*, 703–719.

Thiel, B. L., & Viney, C. (1996). Beta sheets and spider silk. *Science, 273*, 1480–1481.

Tillinghast, E. K., Chase, S. F., & Townley, M. A. (1984). Water extraction by the major ampullate duct during silk formation in the spider, *Argiope aurantia* Lucas. *Journal of Insect Physiology, 30*, 591–596.

Tillinghast, E. K., & Townley, M. A. (1994). Silk glands of araneid spiders – selected morphological and physiological aspects. *Silk Polymers, 544*, 29–44.

Urry, D. W., Haynes, B., Zhang, H., Harris, R. D., & Prasad, K. U. (1988). Mechanochemical coupling in synthetic polypeptides by modulation of an inverse temperature transition. *Proceedings of the National Academy of Sciences of the United States of America, 85*, 3407–3411.

van Beek, J. D., Hess, S., Vollrath, F., & Meier, B. H. (2002). The molecular structure of spider dragline silk: folding and orientation of the protein backbone. *Proceedings of the National Academy of Sciences of the United States of America, 99*, 10266–10271.

Vasanthavada, K., Hu, X., Falick, A. M., La Mattina, C., Moore, A. M. F., Jones, P. R., et al. (2007). Aciniform spidroin, a constituent of egg case sacs and wrapping silk fibers from the black widow spider *Latrodectus hesperus*. *Journal of Biological Chemistry, 282*, 35088–35097.

Viney, C. (1997). Natural silks: archetypal supramolecular assembly of polymer fibres. *Supramolecular Science, 4*, 75–81.

Vollrath, F. (1994). General properties of some spider silks. In D. Kaplan, W. W. Adams, B. Farmer & C. Viney (Eds.), *Silk polymers – materials science and biotechnology* (Vol. 544, pp. 17–28). Washington: ACS.

Vollrath, F. (1999). Biology of spider silk. *International Journal of Biological Macromolecules, 24*, 81–88.

Vollrath, F. (2006). Spider silk: thousands of nano-filaments and dollops of sticky glue. *Current Biology, 16*, R925–R927.

Vollrath, F., & Porter, D. (2006). Spider silk as archetypal protein elastomer. *Soft Matter, 2*, 377–385.

Vollrath, F., & Tillinghast, E. K. (1991). Glycoprotein glue beneath a spider webs aqueous coat. *Naturwissenschaften, 78*, 557–559.

Wen, H. X., Lan, X. Q., Zhang, Y. S., Zhao, T. F., Wang, Y. J., Kajiura, Z., et al. (2010). Transgenic silkworms (*Bombyx mori*) produce recombinant spider dragline silk in cocoons. *Molecular Biology Reports, 37*, 1815–1821.

Werten, M. W., Moers, A. P., Vong, T., Zuilhof, H., van Hest, J. C., & de Wolf, F. A. (2008). Biosynthesis of an amphiphilic silk-like polymer. *Biomacromolecules, 9*, 1705–1711.

Widmaier, D. M., Tullman-Ercek, D., Mirsky, E. A., Hill, R., Govindarajan, S., Minshull, J., et al. (2009). Engineering the *Salmonella* type III secretion system to export spider silk monomers. *Molecular Systems Biology, 5*, 309.

Widmaier, D. M., & Voigt, C. A. (2010). Quantification of the physiochemical constraints on the export of spider silk proteins by *Salmonella* type III secretion. *Microbial Cell Factories, 9*, 78.

Winkler, S., & Kaplan, D. L. (2000). Molecular biology of spider silk. *Journal of Biotechnology, 74*, 85–93.

Winkler, S., Szela, S., Avtges, P., Valluzzi, R., Kirschner, D. A., & Kaplan, D. (1999). Designing recombinant spider silk proteins to control assembly. *International Journal of Biological Macromolecules, 24*, 265–270.

Winkler, S., Wilson, D., & Kaplan, D. L. (2000). Controlling beta-sheet assembly in genetically engineered silk by enzymatic phosphorylation/dephosphorylation. *Biochemistry, 39*, 12739–12746.

Wohlrab, S., Muller, S., Schmidt, A., Neubauer, S., Kessler, H., Leal-Egana, A., et al. (2012). Cell adhesion and proliferation on RGD-modified recombinant spider silk proteins. *Biomaterials, 33*, 6650–6659.

Wong Po Foo, C., Patwardhan, S. V., Belton, D. J., Kitchel, B., Anastasiades, D., Huang, J., et al. (2006). Novel nanocomposites from spider silk–silica fusion (chimeric) proteins. *Proceedings of the National Academy of Sciences of the United States of America, 103*, 9428–9433.

Xia, X.-X., Qian, Z. -G., Ki, C. S., Park, Y. H., Kaplan, D. L., & Lee, S.Y. (2010). Native-sized recombinant spider silk protein produced in metabolically engineered *Escherichia coli* results in a strong fiber. *Proceedings of the National Academy of Sciences, 107*, 14059–14063.

Xie, F., Zhang, H. H., Shao, H. L., & Hu, X. C. (2006). Effect of shearing on formation of silk fibers from regenerated *Bombyx mori* silk fibroin aqueous solution. *International Journal of Biological Macromolecules, 38*, 284–288.

Xu, H. T., Fan, B. L., Yu, S. Y., Huang, Y. H., Zhao, Z. H., Lian, Z. X., et al. (2007). Construct synthetic gene encoding artificial spider dragline silk protein and its expression in milk of transgenic mice. *Animal Biotechnology, 18*, 1–12.

Yamada, H., Nakao, H., Takasu, Y., & Tsubouchi, K. (2001). Preparation of undegraded native molecular fibroin solution from silkworm cocoons. *Materials Science & Engineering. C Biomimetic and Supramolecular Systems, 14*, 41–46.

Yang, M., & Asakura, T. (2005). Design, expression and solid-state NMR characterization of silk-like materials constructed from sequences of spider silk, *Samia cynthia ricini* and *Bombyx mori* silk fibroins. *Journal of Biochemistry, 137*, 721–729.

Yang, J. J., Barr, L. A., Fahnestock, S. R., & Liu, Z. B. (2005). High yield recombinant silk-like protein production in transgenic plants through protein targeting. *Transgenic Research, 14*, 313–324.

Yao, J. M., Masuda, H., Zhao, C. H., & Asakura, T. (2002). Artificial spinning and characterization of silk fiber from *Bombyx mori* silk fibroin in hexafluoroacetone hydrate. *Macromolecules, 35*, 6–9.

Zarkoob, S., Eby, R. K., Reneker, D. H., Hudson, S. D., Ertley, D., & Adams, W. W. (2004). Structure and morphology of electrospun silk nanofibers. *Polymer, 45*, 3973–3977.

Zhang, Y., Hu, J., Miao, Y., Zhao, A., Zhao, T., Wu, D., et al. (2008). Expression of EGFP-spider dragline silk fusion protein in BmN cells and larvae of silkworm showed the solubility is primary limit for dragline proteins yield. *Molecular Biology Reports, 35*, 329–335.

Zhao, A. C., Zhao, T. F., Nakagaki, K., Zhang, Y. S., SiMa, Y. H., Miao, Y. G., et al. (2006). Novel molecular and mechanical properties of egg case silk from wasp spider, *Argiope bruennichi*. *Biochemistry, 45*, 3348–3356.

Zhou, Y. T., Wu, S. X., & Conticello, V. P. (2001). Genetically directed synthesis and spectroscopic analysis of a protein polymer derived from a flagelliform silk sequence. *Biomacromolecules, 2*, 111–125.

Zuo, B., Dai, L., & Wu, Z. (2006). Analysis of structure and properties of biodegradable regenerated silk fibroin fibers. *Journal of Materials Science, 41*, 3357–3361.

Mechanisms of Immune Evasion in Leishmaniasis

Gaurav Gupta*,‡, Steve Oghumu*,,‡, Abhay R. Satoskar*,†,1**
*Department of Pathology, The Ohio State University Medical Center, Columbus, OH, USA
**Department of Oral Biology, The Ohio State University College of Dentistry, Columbus, OH, USA
†Department of Microbiology, The Ohio State University, Columbus, OH, USA
‡These authors contributed equally to this work.
1Corresponding author: E-mail: abhay.satoskar@osumc.edu

Contents

Abstract

Diseases caused by *Leishmania* present a worldwide problem, and current therapeutic approaches are unable to achieve a sterile cure. *Leishmania* is able to persist in host cells by evading or exploiting host immune mechanisms. A thorough understanding of these mechanisms could lead to better strategies for effective management of *Leishmania* infections. Current research has focused on parasite modification of host cell signaling pathways, entry into phagocytic cells, and modulation of cytokine and chemokine profiles that alter immune cell activation and trafficking to sites of infection. Immuno-therapeutic approaches that target these mechanisms of immune evasion by *Leishmania* offer promising areas for preclinical and clinical research.

Diseases caused by *Leishmania* are a major global health problem as over 12 million people currently suffer from leishmaniasis, with an incidence rate of approximately 2 million annually, according to recent estimates (www.who.int/tdr). It is transmitted by sandflies and presents a wide range of clinical manifestations that depend on the specie and strain of *Leishmania*, degree of virulence, and the immunological state of the host. Cutaneous leishmaniasis (CL) manifests as localized skin lesions that may resolve but can become chronic, leading to severe tissue destruction and disfigurement. Lesions could disseminate in immunocompromised patients giving rise to the diffuse cutaneous leishmaniasis (DCL) (Desjeux, 2004). This disease is caused by *Leishmania major* (the Middle East and Mediterranean Region), *Leishmania mexicana* (Central America), and *Leishmania amazonensis* (South America). *Leishmania tropica* and *Leishmania aethiopica* also cause CL in the Old World. Mucocutaneous leishmaniasis caused by *Leishmania braziliensis* is endemic in South America and is clinically characterized by the involvement of the nasal and oropharyngeal mucosa with extensive tissue destruction due to inflammation. Visceral leishmaniasis (VL), the most severe form of leish-maniasis, is caused by *Leishmania donovani* and *Leishmania chagasi* in the Old and New worlds, respectively. Clinical manifestations include hepatomegaly and splenomegaly due to parasite infiltration of the liver and spleen and if left untreated, it is almost always fatal (www.who.int/tdr) (Alexander, Satos-kar, & Russell, 1999; Awasthi, Mathur, & Saha, 2004; Desjeux, 2004).

Chemotherapeutic and prophylactic approaches to the management of the various forms of leishmaniasis have been problematic due to the ten-dency of the parasite to persist within host cells and subsequent failure to achieve a sterile cure. Such latent or chronic forms of the disease present obvious dangers when the hosts' immune system is compromised. The key to effective management of *Leishmania* infections therefore depends on a thorough understanding of the immunological basis for parasite persistence within host cells. Interestingly, macrophages which are the primary immune cells involved in the eradication of *Leishmania* in a mammalian host are the targets for the parasite. How *Leishmania* is able to survive and thrive in this hostile environment as well as capitalize on host defense mechanisms to favor the establishment of disease will be addressed in this review.

1. IMMUNITY TO *LEISHMANIA*

Successful elimination of *Leishmania* depends on the coordinated action of various players of the immune system. From promastigote entry into the blood stream after a sandfly bite to its final mammalian cellular tar-get as an amastigote, the battle between disease establishment and parasite eradication will partly be decided by the ability of *Leishmania* to evade host immunity. Components of host defense important for the eradication of *Leishmania* include both elements of the innate and adaptive immune system.

1.1. The Complement System

It has long been known that the complement system plays a significant role in the eradication of promastigotes in the blood stream of an infected host (Rezai, Sher, & Gettner, 1969). Recent studies have shown that susceptibility of *Leishmania* parasites to complement-mediated lysis in vitro is directly related to the concentration of serum complement (Moreno et al., 2007). Although the classical pathway of the complement system is activated by *Leishmania*, complement-mediated destruction of parasites is generally amplified by the alternate pathway (Hoover, Berger, Nacy, Hockmeyer, & Meltzer, 1984). Amastigotes from *L. donovani* are more resistant than *L. tropica* amastigotes to lysis by complement, suggesting that some *Leishmania* spp. might actively resist this process (Hoover et al., 1984). It should be noted that while unfractionated promastigotes are highly susceptible to lysis by complement, infective metacyclic promastigotes are generally more resistant (Puentes, Sacks, da Silva, & Joiner, 1988).

1.2. Cells of Innate Immunity

Neutrophils are the first host cells to reach the site of *Leishmania* infection within a few hours of inoculation by a sandfly bite (Müller et al., 2001; Pompeu, Freitas, Santos, Khouri, & Barral-Netto, 1991). They have been shown to actively engulf *Leishmania* promastigotes (Mollinedo, Janssen, de la Iglesia-Vicente, Villa-Pulgarin, & Calafat, 2010; Pearson & Steigbigel, 1981) and produce an array of microbicidal factors against *Leishmania* such as nitric oxide (Charmoy et al., 2007), neutrophil elastase (NE) (Ribeiro-Gomes et al., 2007), platelet activating factor (Camussi, Bussolino, Salvidio, & Baglioni, 1987), and neutrophil extracellular traps (Guimarães-Costa et al., 2009). Neutrophils generally have a protective role in most forms of *Leishmania* infections (de Souza Carmo, Katz, & Barbiéri, 2010; Novais et al., 2009), although outcomes are dependent on the *Leishmania* strain, the genetic background of the host and the apoptotic or necrotic state of the neutrophils (Afonso et al., 2008; de Souza Carmo et al., 2010; Filardy et al., 2010; Novais et al., 2009; Ribeiro-Gomes et al., 2004). In vivo depletion of neutrophils during *L. major* infection results in an increase in parasite load in resistant mice (C57BL/6 and C3H/HeJ mice), but susceptible mice (BALB/c mice) show a reduction in parasite load (Ribeiro-Gomes et al., 2004; Tacchini-Cottier et al., 2000).

Early on during *Leishmania* infection, natural killer (NK) cells are also recruited to the infected site after neutrophil recruitment (Müller et al., 2001;Thalhofer, Chen, Sudan, Love-Homan, & Wilson, 2011).They are the primary source of early IFN-γ that favors the Th1 differentiation of CD4$^+$ T cells (Scharton & Scott, 1993) and restricts early parasite dissemination (Diefenbach et al., 1998; Laskay, Diefenbach, Röllinghoff, & Solbach, 1995). NK cells can also mediate direct parasite lysis through its cytotoxic activity (Lieke et al., 2011) and subsequently contribute to cytokine-mediated inducible nitric oxide synthase (iNOS) induction in *Leishmania*-infected macrophages (Prajeeth, Haeberlein, Sebald, Schleicher, & Bogdan, 2011). Studies using NK cell deficient mice have shown the importance of these cells in the containment of *L. donovani* infection but they are dispensable in models of cutaneous leishmaniasis (Kirkpatrick & Farrell, 1982; Satoskar et al., 1999). In humans, NK cell population has been shown to decrease in cases of progressive disease (Pereira et al., 2009; Peruhype-Magalhães et al., 2005), while sites of healing lesions show an infiltration of CD56$^+$ NK cells (Pereira et al., 2009), which suggests a protective role for NK cell population in human leishmaniasis.

Natural killer T (NKT) cells, a specialized subset of T lymphocytes involved in innate immunity to pathogens, also play important roles in the immune response during the early stages of *Leishmania* infection (Amprey et al., 2004; Ishikawa et al., 2000), although these roles vary depending on the specie of *Leishmania* and strain of mice. *Leishmania* surface glycoconjugates and glycoinositol are recognized by CD1d restricted NKT cells (Amprey et al., 2004; Mattner, Donhauser, Werner-Felmayer, & Bogdan, 2006) which contribute to hepatic clearance of *L. donovani* in BALB/c mice through the induction of IFN-γ (Amprey et al., 2004), but exacerbates the disease in C57BL/6 mice (Stanley et al., 2008). In *L. major* infection, protection by NKT cells seems to be organ specific, where they contribute to parasite control in skin lesions and the spleen but not in the lymph nodes (Mattner et al., 2006). In humans, an increase in iNKT cell frequency in the bone marrow of VL patients is observed as the disease progresses, which decreases following treatment (Rai,Thakur, Seth, & Mitra, 2011).

1.3. Cells of Adaptive Immunity

Lymphocytes are involved in adaptive immune responses to *Leishmania* infection, primarily through the elaboration of cytokines that activate or dampen the antiparasitic activity of macrophages. It is well known that T cells play a major role in immunity to the various forms of *Leishmania*. Generally,

IFN-γ-producing Th1 cells are essential to the resolution of infection with *L. major*, where they induce nitric oxide production in macrophages. On the other hand, susceptibility is associated with the production of cytokines produced by Th2 cells such as IL-4 and IL-13. Indeed, genetically resistant (such as C57BL/6) or susceptible (such as BALB/c) mice to *L. major* infection are characterized by their ability to produce Th1 or Th2 cytokine profiles respectively. This is also the case in visceral leishmaniasis caused by *L. donovani*, in mice and humans where IFN-γ-producing Th1 cells protect against severe disease (Kushawaha, Gupta, Sundar, Sahasrabuddhe, & Dube, 2011). In human cutaneous leishmaniasis IFN-γ production by CD8+ T cells seems to contribute to disease resolution (Mary, Auriault, Faugère, & Dessein, 1999; Nateghi Rostami et al., 2010). However, as demonstrated by *Leishmania* infection models using CD4 or CD8 deficient mice, while CD4 T cells are required for resolution of disease (Chakkalath et al., 1995), the role of CD8 T cells in immunity to cutaneous or visceral leishmaniasis seems to depend on the model used (Belkaid, Von Stebut, et al., 2002; Erb, Blank, Ritter, Bluethmann, & Moll, 1996; Gomes-Pereira, Rodrigues, Rolão, Almeida, & Santos-Gomes, 2004; Huber, Timms, Mak, Röllinghoff, & Lohoff, 1998; Tsagozis, Karagouni, & Dotsika, 2005).

Studies using mouse models that affect the activity and migration of regulatory T cells show that they play key roles in regulating the immune response to *L. major* infection (Liu et al., 2009; Suffia, Reckling, Salay, & Belkaid, 2005; Yurchenko et al., 2006). On the one hand, they dampen the effector response of CD4+ T cells partly mediated through IL-10 expression by Tregs. This results in parasite persistence even in resistant C57BL/6 mice, which has implications in the case of disease reactivation. On the other hand, by allowing parasite persistence, Tregs contribute to the maintenance of long term immunity to *L. major* (Belkaid, 2003; Belkaid, Piccirillo, Mendez, Shevach, & Sacks, 2002). This balance between effector and regulatory T cells could potentially be exploited by *Leishmania* parasites to evade host immune responses.

Other immune cells are also involved in controlling immunity to *Leishmania*. B cells seem to play a detrimental role in the early stages of *L. donovani* infection, as shown by significantly reduced parasite burdens in B cell deficient mice (Smelt, Cotterell, Engwerda, & Kaye, 2000). Marginal zone B cells in particular have been shown to dampen the cytotoxic activity of antigen specific CD8 T cells as well as the frequency of IFN gamma producing CD4+ T cells (Bankoti, Gupta, Levchenko, & Stäger, 2012), thus contributing to increased parasite loads. Recent studies on myeloid-derived suppressors cells show that they contribute to resistance to *L. major* in an NO dependent manner, despite their ability to suppress T cell activation (Pereira et al., 2011). During *L. major* infection, dendritic cell subsets as well as their differentiation state play diverse roles in modulating the adaptive immune and affecting the outcome of the disease (Kautz-Neu et al., 2011; Wiethe et al., 2008).

1.4. Cytokines and Chemokines

As shown by a variety of cytokine and chemokine knockout mouse models, cytokines and chemokines play a huge role in immunity to a host of infectious diseases including *Leishmania* infections. While some of these secreted protein immune-modulators are involved in the activation and differentiation of immune cells important in parasite clearance (such as IL-12, TNF-α, IFN-γ) (Mattner et al., 1996; Swihart et al., 1995), others could either dampen the immune response against *Leishmania* or activate and differentiate immune cells that will ultimately favor the persistence of the parasite (such as IL-4, IL-10, IL-13, TGF-β) (Kopf et al., 1996). Cytokines are produced by immune or infected cells and exert their function by activating other cells to release molecules that inhibit or favor the growth of *Leishmania*. Chemokines and chemokine receptors are involved in trafficking of immune cells to inflammatory sites. Indeed, many researchers have suggested their use in the immuno-prophylaxis or therapy of leishmaniasis (Gupta, Majumdar, et al., 2011).

2. MECHANISMS OF IMMUNE EVASION

Like any successful pathogen, *Leishmania* has developed strategies to evade host immune mechanisms in order to survive within the host. A significant number of virulence factors discovered in *Leishmania* are directed against circumventing the host immune response. The ability of *Leishmania* to maintain a chronic infectious state within its host depends to a large extent on its immune evasion potential. Indeed, the ongoing battle between the robust immune response mounted by a host and the counter evasion strategies by the parasite will ultimately decide the fate of the disease. The mechanisms of immune evasion by *Leishmania* species include the following:

2.1. Modification of the Complement System and Phagocytosis

Once the female sandfly injects *Leishmania* promastigotes into the mammalian host, it becomes imperative for the parasite to escape or deactivate the host complement system before entering in the macrophages. *Leishmania* is able to evade the host's complement system using a variety of mechanisms. Unlike noninfective procyclic promastigotes, infective metacyclic promastigotes prevent the insertion of the C5–C9 membrane attack complex making them highly resistant to complement-mediated lysis (Puentes, Da Silva, Sacks, Hammer, & Joiner, 1990). This parasite stage-specific difference in complement evasion is due to a developmental modification resulting in the elongation of lipophosphoglycan (LPG) in metacyclic *Leishmania* (Sacks, Pimenta, McConville, Schneider, & Turco, 1995). Among the various stages of *Leishmania*, metacyclic promastigotes show the highest expression of protein kinases, which phosphorylate complement proteins C3, C5, and C9, resulting in the deactivation of classical and alternative complement pathways (Hermoso, Fishelson, Becker, Hirschberg, & Jaffe, 1991). The activity of *Leishmania* glycoprotein 63 (GP63), a metalloproteinase found abundantly on the surface of metacyclic promastigotes, is pivotal for resisting complement lysis. GP63 cleaves the C3b to an inactive form, C3bi on the surface membrane of the parasite, thereby hindering the formation of C5 convertase (Brittingham et al., 1995).

C3bi also serves as an opsonin, facilitating the uptake of the parasite by binding to complement receptor 3 (CR3) and transiently to CR1 on the surface of macrophages (Kane & Mosser, 2000; Mosser & Edelson, 1985). Attachment via CR3 rather than CR1 is advantageous to the parasite as CR3 ligation even in the absence of *Leishmania* inhibits the production of IL-12 (Marth & Kelsall, 1997). Thus *Leishmania* is able to exploit CR3-mediated phagocytosis, facilitating a 'silent entry' into macrophages thereby evading the host's protective immune response (Da Silva, Hall, Joiner, & Sacks, 1989; Wright & Silverstein, 1983). To further support the role of CR3 in mediating host susceptibility to *Leishmania*, a recent study using CR3 deficient BALB/c mice showed an increased resistance to *L. major* infection (Carter, Whitcomb, Campbell, Mukbel, & McDowell, 2009).

Although macrophages are involved in the eradication of *Leishmania*, they are the primary host cells targeted by the parasite for survival and proliferation. As such *Leishmania* possess a variety of cell surface molecules that enable them localize to macrophages. LPG on the surface of *Leishmania* promastigotes can facilitate binding to mannosyl/fucosyl receptors (Blackwell et al., 1985), complement reactive protein (Culley, Harris, Kaye, McAdam, & Raynes, 1996), and CR4 (Talamás-Rohana, Wright, Lennartz, & Russell, 1990) on macrophage membranes. Furthermore, *Leishmania* glycosylinositol phospholipid (GIPL) and GP63 have important roles in the attachment of parasites to macrophages (Brittingham et al., 1995; Suzuki, Tanaka, Toledo, Takahashi, & Straus, 2002). *Leishmania* amastigotes, unlike promastigotes, can be internalized using phosphatidyl serine and Fc receptors although additional receptors may also participate in the entry process (de Freitas Balanco et al., 2001; Guy & Belosevic, 1993; Weingartner et al., 2012). Recently, our group demonstrated that the phosphoinositide 3-kinase gamma (PI3-Kγ) signaling pathway is exploited by *L. mexicana* to facilitate parasite entry into macrophages and progression of disease. Further, therapy using the PI3-Kγ specific inhibitor AS–605240 protects against cutaneous leishmaniasis caused by *L. mexicana* (Cummings et al., 2012).

2.2. Alteration of Toll-Like Receptor Pathways

Toll-like receptors (TLRs) expressed on the cells of the innate immune system are critical for recognition of pathogen–associated molecular patterns. Initial interaction of the parasite with different TLR's dictates the outcome of the infection (Faria, Reis, & Lima, 2012). TLR2 on host macrophages recognizes *Leishmania* promastigote-derived LPG (de Veer et al., 2003; Flandin, Chano, & Descoteaux, 2006) and amastigote-specific antigens (Srivastava et al., 2012). LPG interaction with TLR2 results in the induction of TNF-α (Flandin et al., 2006), IL-12 (Kavoosi, Ardestani, & Kariminia, 2009), NO (Kavoosi, Ardestani, Kariminia, & Alimohammadian, 2010), and reactive oxygen species (ROS) (Kavoosi et al., 2009). TLR2 agonists which activate this TLR2 signaling pathway has been utilized in inducing a host protective immune response resulting in parasite clearance from *L. donovani* infected macrophages (Bhattacharya et al., 2010).

To subvert this inflammatory response *L. major* recruits suppressors of the cytokine signaling family proteins, SOCS-1 and SOCS-3, which negatively regulates TLR2 induced cytokine induction (de Veer et al., 2003). Another mechanism of TLR2-mediated suppression by *Leishmania* involves the activation of host de-ubiquitinating enzyme A20 by *L. donovani* promastigotes resulting in the impairment of TLR2-mediated release of IL-12 and TNF-α as the parasite interferes with the ubiquitination of TRAF6 (Srivastav, Kar, Chande, Mukhopadhyaya, & Das, 2012). On the other hand, *L. amazonensis* capitalize on TLR2 signaling to facilitate the establishment of infection, by increasing the expression of double stranded RNA dependent protein kinase (PKR) and IFN-β. IFN-β increases superoxide dismutase 1 levels which inhibits superoxide-dependent parasite killing and augments parasite replication (Khouri et al., 2009; Vivarini et al., 2011). Furthermore, a similar study with TLR2-deficient mice showed an essential role of TLR2 in lesion development during *L. braziliensis* infection (Vargas-Inchaustegui et al., 2009). These findings suggest multiple mechanisms employed by different species of *Leishmania* to alter TLR2 signaling and enhance parasite establishment.

TLR4 plays a critical role in shaping the host immune response during *Leishmania* infection. Studies with TLR4$^{-/-}$ mice demonstrate the role of TLR4 in the control of *L. major* infection (Kropf, Freudenberg, Modolell, et al., 2004, Kropf, Freudenberg, Kalis, et al., 2004). Glycosphingophospholipid (GSPL) and proteoglycolipid complex (P8GLC) from *Leishmania* induce TLR4, promoting a strong antiparasitic immune response (Karmakar, Bhaumik, Paul, & De, 2012; Whitaker, Colmenares, Pestana, & McMahon-Pratt, 2008). There is a strong TNF-α response upon P8GLC-TLR4 engagement in macrophages infected with *Leishmania pifanoi* amastigotes and in vivo treatment with P8GLC showed enhanced parasite clearance in TLR4-competent infected mice compared to TLR4$^{-/-}$-deficient infected mice (Whitaker et al., 2008). Similar results were obtained using GSPL treatment in *L. donovani* infection (Karmakar, Paul, & De, 2011).

Leishmania has devised mechanisms to alter TLR4 signaling pathways to favor establishment of infection. During *L. donovani* infection, TLR4-mediated macrophage activation is suppressed through the release of TGF-β that activates the ubiquitin editing enzyme A20 and Src homology 2 domain phosphotyrosine phosphatase 1 (SHP-1) (Das et al., 2012). *Leishmania major* utilizes its inhibitors of serine protease (ISP) to prevent NE-mediated TLR4 activation, inhibiting *Leishmania* uptake and killing by host macrophages (Faria et al., 2011; Ribeiro-Gomes et al., 2007). On the other hand, *L. mexicana* capitalizes on TLR4 signaling to inhibit the production of IL-12 by infected macrophages and promotes parasite establishment (Shweash et al., 2011).

Other TLRs involved in infection with *Leishmania* include TLR3 and TLR9. Endogenous TLR3 binds to double stranded RNA and has been shown to be activated during *Leishmania* infection. A virulent strain of *Leishmania guyanensis* and *Leishmania vianna* which harbor *Leishmania* RNA virus (LRV1) has been shown to activate the TLR3-TRIF–dependent pathway essential for increased pro-inflammatory mediator expression after macrophage infection (Hartley, Ronet, Zangger, Beverley, & Fasel, 2012). Interestingly, TLR3-mediated immune responses rendered mice more susceptible to infection showing increased footpad swelling and parasitemia along with increased metastasis (Ives et al., 2011).

2.3. Surviving in the Phagosome

To survive inside the macrophage, *Leishmania* must resist the harsh conditions created by the phagocyte, including acidic pH, elevated temperature, and increased oxidative/nitrosative stress. Upon phagocytosis, *Leishmania* promastigotes are internalized into endosomal compartments, where they transform into amastigotes. In order to escape the hostile environment of the phagolysosome, promastigotes transiently prevent fusion of the phagosome and lysosome thereby delaying or inhibiting endosomal maturation as observed by the late expression of Rab7 and LAMP-1 (Olivier, Gregory, & Forget, 2005; Scianimanico et al., 1999). LPG on *Leishmania* promastigotes inhibits endosome maturation by inducing periphagosomal F-actin accumulation (Holm, Tejle, Magnusson, Descoteaux, & Rasmusson, 2001). LPG also prevents acidification of the phagosome by interfering with the V-ATPase pump, which allows promastigotes to differentiate into resistant amastigotes (Vinet, Fukuda, Turco, & Descoteaux, 2009). Scavenging on host sphingolipids is another survival strategy employed by *Leishmania* amastigotes for counteracting the acidic environment of the phagolysosome (Ali, Harding, & Denny, 2012). In order to dilute the leishmanicidal effect of nitric oxide, *Leishmania* regulates the lysosomal trafficking (LYST) protein resulting in the formation of large parasitophorous vacuoles (Wilson et al., 2008).

Leishmania amastigotes require iron for metabolism and replication, therefore acquisition of iron is critical for its survival within the macrophage phagolysosome (Huynh & Andrews, 2008). In murine macrophages, Nramp1 functions as an efflux pump that translocates Fe^{2+} from the phagolysosome into the cytosol thereby restricting its availability to the parasite. *Leishmania* counteracts this effect by the activation of its own iron transporters, LIT1 and LIT2 which effectively compete with the host's iron sequestering mechanism (Kaye & Scott, 2011). Arginine is also an essential growth factor required by intracellular *Leishmania* amastigotes for the synthesis of polyamines, but is also used by macrophages for microbicidal nitric oxide production (Iniesta, Gómez-Nieto, & Corraliza, 2001; Kropf, Herath, Weber, Modolell, & Müller, 2003). As demonstrated by studies using gene deficient mutants of *Leishmania*, the parasite encoded arginase enzyme not only facilitates growth by supplying essential nutrients to the parasite through the polyamine pathway, but also attenuates the iNOS-dependent killing mechanism of infected macrophages by competing for available intracellular arginine (Gaur et al., 2007; Reguera, Balaña-Fouce, Showalter, Hickerson, & Beverley, 2009). These virulence factors expressed by *Leishmania* amastigotes enable the parasite to evade the host's cellular immune response and survive in the phagosome.

2.4. Defective Antigen Presentation and Co-stimulation

Virulent stages of *Leishmania* have the capacity to attenuate T cell–mediated immune responses by regulating the expression of leishmanial antigen loaded major histocompatibility complex (MHC) molecules on antigen presenting cells. They accomplish this by antigen sequestration or interference with the loading of antigens onto MHC class II molecules (Fruth, Solioz, & Louis, 1993; Kima, Soong, Chicharro, Ruddle, & McMahon-Pratt, 1996; Prina, Lang, Glaichenhaus, & Antoine, 1996). Studies on presentation of the *Leishmania* antigen, *Leishmania* homolog of receptors for activated C kinase (LACK), by infected macrophages showed a transient yet strong LACK-specific T cell response when infected with stationary or log phase *Leishmania* promastigotes. On the other hand, murine macrophages infected with either *Leishmania* metacyclic promastigotes or amastigotes showed a weak or absent LACK-specific T cell activation respectively (Courret et al., 1999). Membrane lipid rafts are important platforms for antigen presentation as it concentrates MHC class II molecules into microdomains, which allow efficient antigen presentation at low peptide densities. *Leishmania donovani* increases the fluidity of lipid rafts in macrophages resulting in defective antigen presentation and diminished T cells responses (Chakraborty et al., 2005). MHC class II, but not class I molecules have been shown to be important in host resistance against *Leishmania* (Huber et al., 1998; Locksley, Reiner, Hatam, Littman, & Killeen, 1993), and only MHC II molecules appear in the parasitophorous vacuole of an infected macrophage (Lang et al., 1994). MHC II molecules, located in specific organelles called megasomes, are endocytosed by *Leishmania* amastigotes and degraded by cysteine proteases (De Souza Leao, Lang, Prina, Hellio, & Antoine, 1995). These examples highlight the ability of *Leishmania* to evade recognition by T lymphocytes by interfering with macrophage antigen presentation.

Co-stimulatory molecules such as B7-1, B7-2, and CD40 expressed on macrophages are also critical for setting up antiparasitic T cell responses (Alexander et al., 1999; Bogdan, Gessner, Solbach, & Röllinghoff, 1996). In *Leishmania*-infected macrophages, there is a reduced expression of B7-1 even after lipopolysaccharide (LPS) stimulation, possibly mediated by parasite-induced prostaglandins (Saha, Das,Vohra, Ganguly, & Mishra, 1995). Similarly, during *L. major* infection B7-1 expression is down-regulated in epidermal cells from susceptible mice (Mbow, DeKrey, & Titus, 2001). Indirect evidence seems to support the notion that *Leishmania* increases the expression of B7-2 to favor the establishment of disease. In the same study by Mbow et al. (2001) B7-2 expression was higher in epidermal cells from susceptible mice than in resistant mice after *L. major* infection. In another study with human leishmaniasis caused by *L. mexicana*, B7-2 expression was higher in monocytes from patients with active DCL, while no change was observed in the expression of CD40 and B7-1 (Carrada et al., 2007). Furthermore, antibody blockade of B7-2, but not B7-1 led to an increased T cell response resulting in diminished parasite load (Murphy, Cotterell, Gorak, Engwerda, & Kaye, 1998; Murphy, Engwerda, Gorak, & Kaye, 1997). This is further supported by studies with *L. major* infected B7-1 knockout, B7-2 knockout, and B7-1/B7-2 double knockout BALB/c mice (Brown et al., 2002). Signaling via CD40-CD40L interactions, critical for the induction of antileishmanial responses (Campbell et al., 1996; Kamanaka et al., 1996), is also impaired by *L. major* (Awasthi et al., 2003).

2.5. Alteration of Host Cell Signaling

In order to survive inside macrophages armed with a host of microbicidal factors, *Leishmania* interferes with cell signaling cascades involved in their synthesis. One factor critical for early protective response against *Leishmania* (Murray & Nathan, 1999) is reactive oxygen species which is activated by the nicotinamide adenine dinucleotide phosphate (NADPH) oxidase complex (De Leo, Ulman, Davis, Jutila, & Quinn, 1996). In *L. donovani*-infected macrophages, phosphorylation of the p67 and p47 subunits of NADPH oxidase is blocked due to impaired protein kinase C (PKC) signaling (Bhattacharyya, Ghosh, Sen, Roy, & Majumdar, 2001) which favors parasite survival (Olivier, Brownsey, & Reiner, 1992; Turco et al., 1987). *Leishmania* promastigote LPG deactivates PKC by interfering with the binding of Ca^{2+} and diacyl glycerol to PKC and obstructing the insertion of PKC into the membrane (Descoteaux & Turco, 1999). Recent studies in *L. mexicana*-infected macrophages show that the ability of parasite-derived LPG to regulate the oxidative burst via PKC-α activity is related to the susceptible/resistant phenotypes observed genetic strains of mice (Delgado-Domínguez et al., 2010). Interestingly, amastigotes, devoid of LPG, have also shown similar attenuation in PKC activity (Olivier et al., 1992) suggesting the existence of LPG independent mechanisms such as *Leishmania*-induced ceramide (Ghosh et al., 2001) and IL–10 induction (Bhattacharyya, Ghosh, Jhonson, Bhattacharya, & Majumdar, 2001). GP63 which cleaves the PKC substrate, myristoylated alanine rich C kinase (MARCKS) and MARCKS-related protein (Olivier, Atayde, Isnard, Hassani, & Shio, 2012), and GIPL (Chawla & Vishwakarma, 2003) are virulence factors that are involved in the inhibition of PKC activity.

IFN-γ-mediated activation of the infected macrophage is critical for the destruction of intracellular *Leishmania* parasites. IFN-γ signals through the JAK/STAT pathway and is critical for the induction of nitric oxide, but this pathway is suppressed by *Leishmania* (Nandan & Reiner, 1995). Interaction of *Leishmania* promastigotes with macrophages activates SHP-1 which shows a strong and increased interaction with JAK2 resulting in its deactivation (Blanchette, Racette, Faure, Siminovitch, & Olivier, 1999). Further studies show that JAK2 inactivation by SHP-1 results in diminished nitric oxide production which favors parasite survival (Blanchette, Abu–Dayyeh, Hassani, Whitcombe, & Olivier, 2009). Further in *Leishmania* infected macrophages, STAT1 inactivation occurs and translocation of STAT1α to the nucleus is significantly diminished through a mechanism independent of SHP-1 activity (Forget, Gregory, & Olivier, 2005). In both *L. major* and *L. mexicana* infected macrophages treated with IFN-γ, phosphorylation of STAT1α is reduced, but dominant negative STAT1β phosphorylation is increased only in *L. mexicana* infected macrophages which points to species specific differences in the regulation of the JAK/STAT signaling pathway (Bhardwaj, Rosas, Lafuse, & Satoskar, 2005). Furthermore, quenching of membrane cholesterol has recently been shown to be a mechanism by which *Leishmania* impairs with IFN-γ signaling, and re-association of the signaling assembly could be restored by liposomal delivery of cholesterol together with IFN-γ (phospho-JAK1, JAK2, and STAT1) (Sen, Roy, Mukherjee, Mukhopadhyay, & Roy, 2011).

The MAPK signaling pathway (ERK1/2, JNK, and p38MAPK), required for the production of various effector molecules including cytokines and chemokines, is exploited by *Leishmania* as another immune evasive mechanism. This complex signaling pathway could either be activated or suppressed leading to the induction of gene products that favor parasite survival. While *L. donovani* inhibits EKR1/2, p38MAPK, and JNK activation in naïve macrophages leading to the suppression of pro-inflammatory cytokine production (Privé & Descoteaux, 2000), *L. amazonensis* activates ERK1/2 leading to the production of IL-10, a cytokine that contributes to parasite growth (Yang, Mosser, & Zhang, 2007). Mechanisms behind MAPK inactivation by *Leishmania* have been extensively studied. Studies with LPS activated macrophages show that *Leishmania* amastigotes can block the activation of EKR1/2 via the activation of ecto-protein phosphatase (Martiny, Meyer-Fernandes, de Souza, & Vannier-Santos, 1999). In PMA-activated macrophages, *Leishmania* amastigotes deactivated MAP kinase activity (ERK1/2), c-FOS and iNOS expression by activating cellular phosphotyrosine phosphatases (Nandan, Lo, & Reiner, 1999). Studies with SHP-1 deficient infected macrophages confirmed the pivotal role played by this phosphatase in the regulation of ERK1/2 signaling (Forget, Gregory, Whitcombe, & Olivier, 2006). *Leishmania* can also deactivate ERK1/2 by inducing ceramide generation in susceptible host macrophages (Ghosh et al., 2002). Interestingly, *L. mexicana* amastigotes block LPS induced IL-12 production by relying on their own cysteine proteinases for degrading the ERK1/2 and JNK but not p38MAPK (Cameron et al., 2004). Studies with CD40 ligand/antibody, showed that *Leishmania* interferes with the strength of CD40 cross-linking resulting in the reciprocal regulation of ERK1/2 and p38MAPK which governs the production of IL-10 and IL-12 in *Leishmania*-infected macrophages (Mathur, Awasthi, Wadhone, Ramanamurthy, & Saha, 2004). Moreover, *Leishmania* can redirect CD40-regulated immune responses via the reciprocal activation of MAPK phosphatases (MKP), MKP-1 and MKP-3 which reveal a novel parasite-devised immune evasion strategy (Srivastava, Sudan, & Saha, 2011).

Leishmania also regulates other important transcription factors such as AP-1 and NF-kB which have been shown to be important in immunity against the parasite. One study using RelA (NF-kB p65 subunit) knockout infected macrophages show higher intracellular parasite load and reduced NO levels compared to WT (Mise-Omata et al., 2009), establishing the importance of NF-kB in resistance to the disease. *Leishmania* specifically reduces the overall expression of RelA (Calegari-Silva et al., 2009). *Leishmania mexicana* promastigotes rely on the virulence factors GP63 and cysteine peptidase activity to cleave the RelA-p65 subunit into the smaller RelA-p35 subunit which activates specific chemokines that favor parasite multiplication (Abu-Dayyeh, Hassani, Westra, Mottram, & Olivier, 2010; Gregory, Godbout, Contreras, Forget, & Olivier, 2008). *Leishmania mexicana* amastigotes degrade the RelA-p65 subunit though cellular tyrosine phosphatases upon activation by parasite's cysteine peptidase (Abu-Dayyeh et al., 2010). *Leishmania major* amastigotes favor IL-10 induction by selectively inhibiting p65–p50 complex thus allowing selective nuclear translocation of p50–p35 complex (Guizani-Tabbane, Ben-Aissa, Belghith, Sassi, & Dellagi,

2004). *Leishmania* also attenuates the activity of AP-1 (composed of dimmers of Fos and Jun family members) in infected macrophages in order to regulate the production of pro-inflammatory cytokines IL-1β, TNF-α, and IL-12 (Abu-Dayyeh et al., 2008; Contreras et al., 2010). Previous studies have shown that *Leishmania* interferes with the nuclear translocation of AP-1 which might be due to parasite-induced ceramide generation (Ghosh et al., 2002). Recent reports further show that *Leishmania*-derived GP63 enters the nucleus of the host macrophage to cleave c-Jun and c-Fos subunits of AP-1 rendering them inactive (Contreras et al., 2010). Thus by using multiple mechanisms *Leishmania* is able to interfere with the activity of transcription factors thereby subverting the host immune response.

Other host signaling targets of *Leishmania* include the mammalian mechanistic target of rapamycin (mTOR), which is cleaved by the *Leishmania* protease GP63 leading to the inhibition of mTOR complex 1 (mTORC1) and concomitant activation of 4E-BP1 to promote parasite survival (Jaramillo et al., 2011). Recently, *L. donovani* has been shown to utilize the mTOR signaling pathway to regulate IL-12 and IL-10 production in infected macrophages (Cheekatla, Aggarwal, & Naik, 2012).

2.6. Modulation of Cytokines and Chemokines

By interfering with host cell signaling pathways, many of which are still yet to be completely understood, various species of *Leishmania* possess the capability to modulate the cytokine profile of infected host cells to favor dissemination of the parasite and prevent its eradication. Pro-inflammatory cytokines (such as IL-12), which are essential for the generation of a successful immune response against *Leishmania*, are generally suppressed, facilitating a 'silent entry' of the parasite into host macrophages (Belkaid et al., 2000; McDowell & Sacks, 1999; Reiner, Zheng, Wang, Stowring, & Locksley, 1994; Weinheber, Wolfram, Harbecke, & Aebischer, 1998). Conversely, IL-10 which is essential for parasite survival and disease progression is induced by *Leishmania* in infected monocytes and macrophages (Chandra & Naik, 2008; Meddeb-Garnaoui, Zrelli, & Dellagi, 2009). Leishmania also expresses cytokine orthologs that modulate host immune cells. The human macrophage migration inhibitory factor ortholog, produced by *L. major*, has been shown to inhibit macrophage apoptosis and could contribute to parasite persistence and evasion of immune destruction (Kamir et al., 2008).

Chemokines and chemokine receptors play a major role in immunity to *Leishmania* by coordinating the recruitment and activation of anti-leishmanial immune cells (Oghumu, Lezama-Dávila, Isaac-Márquez,

& Satoskar, 2010). *Leishmania's* attempt to alter the chemokine expression profile in the infected tissue microenvironment will therefore contribute to their ability to evade the host's immune system. For over a decade, *Leishmania* has been known to induce or inhibit the expression of chemokines to control recruitment of immune cells. In human cutaneous leishmaniasis, the diffuse form of the disease has been associated with lower levels of the macrophage chemotactic factor CCL2 (MCP-1), which is higher in the localized form of the disease (Ritter et al., 1996). Since CCL2 also induces antiparasitic activity in macrophages (Mannheimer, Hariprashad, Stoeckle, & Murray, 1996), inhibition of this chemokine by *Leishmania* could potentially facilitate its survival within the host. The *Leishmania* virulence factor LPG can also inhibit migration of monocytes across the endothelial wall by inhibiting the synthesis of CCL2 and the expression of cell surface adhesion molecules including E-selectin, ICAM-1, and VCAM-1 by endothelial cells (Lo et al., 1998). Expression of chemokine receptors CCR4 and CCR5 as well as integrin VLA-4 activity, all involved in macrophage adhesion, has also been shown to be inhibited during *Leishmania* infection (Pinheiro et al., 2006).

The ability of *Leishmania* to selectively recruit immune cells that will facilitate parasite survival was demonstrated by Katzman and Fowell (2008). This study showed that *L. major* can induce the expression of the Th2 attracting chemokine CCL7 at the dermal infection site, allowing the accumulation of IL-4 producing but not IFN-γ producing effector T cells. In contrast, ovalbumin with complete Freund's adjuvant (OVA/CFA) immunized dermis permitted the accumulation of both populations of effector T cells, suggesting that *L. major* actively modulates the local tissue environment in an attempt to evade the hosts' immune response. *Leishmania* promastigotes themselves secrete a chemotactic factor for PMNs, which serve as the host cell in the very early stages of infection (van Zandbergen, Hermann, Laufs, Solbach, & Laskay, 2002). At the same time they can induce the expression of IL-8 by these cells so as to recruit more PMNs which aid in the establishment of infection and proliferation of the parasite. On the other hand, they have the ability to inhibit the expression of IP-10, a chemokine that recruits and activates NK and Th1 cells, which are important in parasite eradication (van Zandbergen et al., 2002).

In a chronic murine VL model, *L. donovani* inhibited dendritic cell migration to T cell areas of the spleen due to a loss in CCR7 expression. Further, treatment of infected mice by adoptive transfer of CCR7 expressing dendritic cells significantly reduced parasite burdens (Ato, Stäger, Engwerda, & Kaye, 2002). Although mechanisms behind the process are still not fully understood, the induction or suppression of chemokine/chemokine receptor expression is a major way *Leishmania* evades the host immune system to favor establishment of disease.

2.7. Modification of T Cell Responses

Th1 cells play a vital role in the elimination of *Leishmania* through the secretion of IFN-γ and CD40L which activate macrophages to produce nitric oxide, a leishmanicidal factor. Recent studies have shown that some secreted factors from *L. major* possess immunosuppressive properties, could dampen lymphoproliferative capabilities and skew T cell polarization toward a susceptible Th2 phenotype (Tabatabaee, Abolhassani, Mahdavi, Nahrevanian, & Azadmanesh, 2011). Similarly, extracts of *L. amazonensis*, a causative agent of CL in the New World, promote a Th2 type immune response in the host, thereby enhancing infection (Silva et al., 2011).

The ability of *Leishmania* to induce the activity of regulatory T cells that suppress anti-*Leishmania* immune responses has been well established in cutaneous and visceral animal models of leishmaniasis (Belkaid, 2003; Belkaid, Piccirillo, et al., 2002; Mendez, Reckling, Piccirillo, Sacks, & Belkaid, 2004). This has also been shown in humans (Ganguly et al., 2010; Katara, Ansari, Verma, Ramesh, & Salotra, 2011; Rai et al., 2012). Tregs generally function as immune-regulators of cell-mediated immune responses, preventing pathology due to uncontrolled effector T cell activity. While this presents apparent advantages to the host, it could very well be capitalized by *Leishmania*, thereby preventing complete eradication of the parasite by Th1 cells. Tregs are retained at the site of infection where they secrete IL-10 and TGF-β which down regulate Th1 and macrophage activity, rendering the area of infection, an immune privileged site (Peters & Sacks, 2006).

In conclusion, current research that focuses on mechanisms of immune evasion by *Leishmania* could shed light on promising targets for therapeutic intervention. This is especially important as medication currently employed in the therapy of *Leishmania* presents problems due to toxicity, patient compliance, and drug resistance. Recent advances in the use of immuno-therapeutic approaches targeted at circumventing the parasites' attempt at host immune evasion include the use of TLR agonists that induce strong cell mediated immune responses. The TLR 4 agonist monophosphoryl lipid A has been used in combination with *Leishmania* antigens or other agonists to protect against cutaneous or mucocutaneous *Leishmania* infection models (Aebischer et al., 2000; Coler & Reed, 2005; Raman et al., 2010) and some formulations are currently being used successfully in clinical trials

(Llanos-Cuentas et al., 2010). Activation of the TLR2 pathway has been shown to reduce parasite burdens in *L. donovani*-infected macrophages (Bhattacharya et al., 2010). TLR2 agonist Pam3Cys and CPG-ODN 2006 have been used in combination with miltefosine in the immunotherapy of experimental visceral leishmaniasis (Gupta, Sane, Shakya, Vishwakarma, & Haq, 2011; Shakya, Sane, Shankar, & Gupta, 2011).

Other promising immuno-therapeutic approaches involve the use of chemokines or chemokine receptor agonists (Gupta, Majumdar, et al., 2011), blockade of co-stimulatory molecules (Murphy et al., 1998, 1997), reconstitution of membrane cholesterol to facilitate re-association of membrane signaling assemblies (Sen et al., 2011), and inhibitors of cell signaling molecules (Cheekatla et al., 2012; Cummings et al., 2012), all mechanisms of which are regulated by *Leishmania* to evade or exploit host immune responses. While this presents an area for extensive future research, current advances have made significant impact and present weighty implications in the management of leishmaniasis.

ACKNOWLEDGMENTS

This work was supported by National Institutes of Health grants R03AI090231, RC4AI092624, R34AI100789, R21AT004160 and R03CA164399 awarded to A.R.S and National Institute of Dental and Craniofacial Research Training Grant T32DE014320 awarded to S.O.

REFERENCES

Abu-Dayyeh, I., Hassani, K., Westra, E. R., Mottram, J. C., & Olivier, M. (2010). Comparative study of the ability of *Leishmania mexicana* promastigotes and amastigotes to alter macrophage signaling and functions. *Infection and Immunity, 78*, 2438–2445.

Abu-Dayyeh, I., Shio, M. T., Sato, S., Akira, S., Cousineau, B., & Olivier, M. (2008). Leishmania-induced IRAK-1 inactivation is mediated by SHP-1 interacting with an evolutionarily conserved KTIM motif. *PLoS Neglected Tropical Diseases, 2*, e305.

Aebischer, T., Wolfram, M., Patzer, S. I., Ilg, T., Wiese, M., & Overath, P. (2000). Subunit vaccination of mice against new world cutaneous leishmaniasis: comparison of three proteins expressed in amastigotes and six adjuvants. *Infection and Immunity, 68*, 1328–1336.

Afonso, L., Borges, V. M., Cruz, H., Ribeiro-Gomes, F. L., DosReis, G. A., Dutra, A. N., et al. (2008). Interactions with apoptotic but not with necrotic neutrophils increase parasite burden in human macrophages infected with *Leishmania amazonensis. Journal of Leukocyte Biology, 84*, 389–396.

Alexander, J., Satoskar, A. R., & Russell, D. G. (1999). *Leishmania* species: models of intracellular parasitism. *Journal of Cell Science, 112*(Pt 18), 2993–3002.

Ali, H. Z., Harding, C. R., & Denny, P. W. (2012). Endocytosis and sphingolipid scavenging in *Leishmania mexicana* amastigotes. *Biochemistry Research International, 2012*, 691363.

Amprey, J. L., Im, J. S., Turco, S. J., Murray, H. W., Illarionov, P. A., Besra, G. S., et al. (2004). A subset of liver NK T cells is activated during *Leishmania donovani* infection by CD1d-bound lipophosphoglycan. *Journal of Experimental Medicine, 200*, 895–904.

Ato, M., Stäger, S., Engwerda, C. R., & Kaye, P. M. (2002). Defective CCR7 expression on dendritic cells contributes to the development of visceral leishmaniasis. *Nature Immunology, 3*, 1185–1191.

Awasthi, A., Mathur, R., Khan, A., Joshi, B. N., Jain, N., Sawant, S., et al. (2003). CD40 signaling is impaired in *L. major*-infected macrophages and is rescued by a p38MAPK activator establishing a host-protective memory T cell response. *Journal of Experimental Medicine, 197*, 1037–1043.

Awasthi, A., Mathur, R. K., & Saha, B. (2004). Immune response to *Leishmania* infection. *Indian Journal of Medical Research, 119*, 238–258.

Bankoti, R., Gupta, K., Levchenko, A., & Stäger, S. (2012). Marginal zone B cells regulate antigen-specific T cell responses during infection. *Journal of Immunology, 188*, 3961–3971.

Belkaid, Y. (2003). The role of CD4(+)CD25(+) regulatory T cells in *Leishmania* infection. *Expert Opinion on Biological Therapy, 3*, 875–885.

Belkaid, Y., Mendez, S., Lira, R., Kadambi, N., Milon, G., & Sacks, D. (2000). A natural model of *Leishmania major* infection reveals a prolonged "silent" phase of parasite amplification in the skin before the onset of lesion formation and immunity. *Journal of Immunology, 165*, 969–977.

Belkaid, Y., Piccirillo, C. A., Mendez, S., Shevach, E. M., & Sacks, D. L. (2002). CD4+CD25+ regulatory T cells control *Leishmania major* persistence and immunity. *Nature, 420*, 502–507.

Belkaid, Y., Von Stebut, E., Mendez, S., Lira, R., Caler, E., Bertholet, S., et al. (2002). CD8+ T cells are required for primary immunity in C57BL/6 mice following low-dose, intradermal challenge with *Leishmania major*. *Journal of Immunology, 168*, 3992–4000.

Bhardwaj, N., Rosas, L. E., Lafuse, W. P., & Satoskar, A. R. (2005). *Leishmania* inhibits STAT1-mediated IFN-gamma signaling in macrophages: increased tyrosine phosphorylation of dominant negative STAT1beta by *Leishmania mexicana*. *International Journal of Parasitology, 35*, 75–82.

Bhattacharya, P., Bhattacharjee, S., Gupta, G., Majumder, S., Adhikari, A., Mukherjee, A., et al. (2010). Arabinosylated lipoarabinomannan-mediated protection in visceral leishmaniasis through up-regulation of toll-like receptor 2 signaling: an immunoprophylactic approach. *Journal of Infectious Diseases, 202*, 145–155.

Bhattacharyya, S., Ghosh, S., Jhonson, P. L., Bhattacharya, S. K., & Majumdar, S. (2001). Immunomodulatory role of interleukin-10 in visceral leishmaniasis: defective activation of protein kinase C-mediated signal transduction events. *Infection and Immunity, 69*, 1499–1507.

Bhattacharyya, S., Ghosh, S., Sen, P., Roy, S., & Majumdar, S. (2001). Selective impairment of protein kinase C isotypes in murine macrophage by *Leishmania donovani*. *Molecular and Cellular Biochemistry, 216*, 47–57.

Blackwell, J. M., Ezekowitz, R. A., Roberts, M. B., Channon, J. Y., Sim, R. B., & Gordon, S. (1985). Macrophage complement and lectin-like receptors bind *Leishmania* in the absence of serum. *Journal of Experimental Medicine, 162*, 324–331.

Blanchette, J., Abu-Dayyeh, I., Hassani, K., Whitcombe, L., & Olivier, M. (2009). Regulation of macrophage nitric oxide production by the protein tyrosine phosphatase Src homology 2 domain phosphotyrosine phosphatase 1 (SHP-1). *Immunology, 127*, 123–133.

Blanchette, J., Racette, N., Faure, R., Siminovitch, K. A., & Olivier, M. (1999). Leishmania-induced increases in activation of macrophage SHP-1 tyrosine phosphatase are associated with impaired IFN-gamma-triggered JAK2 activation. *European Journal of Immunology, 29*, 3737–3744.

Bogdan, C., Gessner, A., Solbach, W., & Röllinghoff, M. (1996). Invasion, control and persistence of *Leishmania* parasites. *Current Opinion in Immunology, 8*, 517–525.

Brittingham, A., Morrison, C. J., McMaster, W. R., McGwire, B. S., Chang, K. P., & Mosser, D. M. (1995). Role of the *Leishmania* surface protease gp63 in complement fixation, cell adhesion, and resistance to complement-mediated lysis. *Journal of Immunology, 155*, 3102–3111.

Brown, J. A., Greenwald, R. J., Scott, S., Schweitzer, A. N., Satoskar, A. R., Chung, C., et al. (2002). T helper differentiation in resistant and susceptible B7-deficient mice infected with *Leishmania major*. *European Journal of Immunology, 32*, 1764–1772.

Calegari-Silva, T. C., Pereira, R. M., De-Melo, L. D., Saraiva, E. M., Soares, D. C., Bellio, M., et al. (2009). NF-kappaB-mediated repression of iNOS expression in *Leishmania amazonensis* macrophage infection. *Immunology Letters, 127*, 19–26.

Cameron, P., McGachy, A., Anderson, M., Paul, A., Coombs, G. H., Mottram, J. C., et al. (2004). Inhibition of lipopolysaccharide-induced macrophage IL-12 production by *Leishmania mexicana* amastigotes: the role of cysteine peptidases and the NF-kappaB signaling pathway. *Journal of Immunology, 173*, 3297–3304.

Campbell, K. A., Ovendale, P. J., Kennedy, M. K., Fanslow, W. C., Reed, S. G., & Maliszewski, C. R. (1996). CD40 ligand is required for protective cell-mediated immunity to *Leishmania major*. *Immunity, 4*, 283–289.

Camussi, G., Bussolino, F., Salvidio, G., & Baglioni, C. (1987). Tumor necrosis factor/cachectin stimulates peritoneal macrophages, polymorphonuclear neutrophils, and vascular endothelial cells to synthesize and release platelet-activating factor. *Journal of Experimental Medicine, 166*, 1390–1404.

Carrada, G., Cañeda, C., Salaiza, N., Delgado, J., Ruiz, A., Sanchez, B., et al. (2007). Monocyte cytokine and costimulatory molecule expression in patients infected with *Leishmania mexicana*. *Parasite Immunology, 29*, 117–126.

Carter, C. R., Whitcomb, J. P., Campbell, J. A., Mukbel, R. M., & McDowell, M. A. (2009). Complement receptor 3 deficiency influences lesion progression during *Leishmania major* infection in BALB/c mice. *Infection and Immunity, 77*, 5668–5675.

Chakkalath, H. R., Theodos, C. M., Markowitz, J. S., Grusby, M. J., Glimcher, L. H., & Titus, R. G. (1995). Class II major histocompatibility complex-deficient mice initially control an infection with *Leishmania major* but succumb to the disease. *Journal of Infectious Diseases, 171*, 1302–1308.

Chakraborty, D., Banerjee, S., Sen, A., Banerjee, K. K., Das, P., & Roy, S. (2005). *Leishmania donovani* affects antigen presentation of macrophage by disrupting lipid rafts. *Journal of Immunology, 175*, 3214–3224.

Chandra, D., & Naik, S. (2008). *Leishmania donovani* infection down-regulates TLR2-stimulated IL-12p40 and activates IL-10 in cells of macrophage/monocytic lineage by modulating MAPK pathways through a contact-dependent mechanism. *Clinical and Experimental Immunology, 154*, 224–234.

Charmoy, M., Megnekou, R., Allenbach, C., Zweifel, C., Perez, C., Monnat, K., et al. (2007). *Leishmania major* induces distinct neutrophil phenotypes in mice that are resistant or susceptible to infection. *Journal of Leukocyte Biology, 82*, 288–299.

Chawla, M., & Vishwakarma, R. A. (2003). Alkylacylglycerolipid domain of GPI molecules of *Leishmania* is responsible for inhibition of PKC-mediated c-fos expression. *Journal of Lipid Research, 44*, 594–600.

Cheekatla, S. S., Aggarwal, A., & Naik, S. (2012). mTOR signaling pathway regulates the IL-12/IL-10 axis in *Leishmania donovani* infection. *Medical Microbiology and Immunology, 201*, 37–46.

Coler, R. N., & Reed, S. G. (2005). Second-generation vaccines against leishmaniasis. *Trends in Parasitology, 21*, 244–249.

Contreras, I., Gómez, M. A., Nguyen, O., Shio, M. T., McMaster, R. W., & Olivier, M. (2010). *Leishmania*-induced inactivation of the macrophage transcription factor AP-1 is mediated by the parasite metalloprotease GP63. *PLoS Pathogens, 6*, e1001148.

Courret, N., Prina, E., Mougneau, E., Saraiva, E. M., Sacks, D. L., Glaichenhaus, N., et al. (1999). Presentation of the *Leishmania* antigen LACK by infected macrophages is dependent upon the virulence of the phagocytosed parasites. *European Journal of Immunology, 29,* 762–773.

Culley, F. J., Harris, R. A., Kaye, P. M., McAdam, K. P., & Raynes, J. G. (1996). C-reactive protein binds to a novel ligand on *Leishmania donovani* and increases uptake into human macrophages. *Journal of Immunology, 156,* 4691–4696.

Cummings, H. E., Barbi, J., Reville, P., Oghumu, S., Zorko, N., Sarkar, A., et al. (2012). Critical role for phosphoinositide 3-kinase gamma in parasite invasion and disease progression of cutaneous leishmaniasis. *Proceedings of the National Academy of Sciences of the United States of America, 109,* 1251–1256.

Da Silva, R. P., Hall, B. F., Joiner, K. A., & Sacks, D. L. (1989). CR1, the C3b receptor, mediates binding of infective *Leishmania major* metacyclic promastigotes to human macrophages. *Journal of Immunology, 143,* 617–622.

Das, S., Pandey, K., Kumar, A., Sardar, A. H., Purkait, B., Kumar, M., et al. (2012). TGF-β(1) re-programs TLR4 signaling in *L. donovani* infection: enhancement of SHP-1 and ubiquitin-editing enzyme A20. *Immunology and Cell Biology, 90,* 640–654.

De Leo, F. R., Ulman, K. V., Davis, A. R., Jutila, K. L., & Quinn, M. T. (1996). Assembly of the human neutrophil NADPH oxidase involves binding of p67phox and flavocytochrome b to a common functional domain in p47phox. *Journal of Biological Chemistry, 271,* 17013–17020.

De Souza Leao, S., Lang, T., Prina, E., Hellio, R., & Antoine, J. C. (1995). Intracellular *Leishmania amazonensis* amastigotes internalize and degrade MHC class II molecules of their host cells. *Journal of Cell Science, 108*(Pt 10), 3219–3231.

Delgado-Domínguez, J., González-Aguilar, H., Aguirre-García, M., Gutiérrez-Kobeh, L., Berzunza-Cruz, M., Ruiz-Remigio, A., et al. (2010). *Leishmania mexicana* lipophosphoglycan differentially regulates PKCalpha-induced oxidative burst in macrophages of BALB/c and C57BL/6 mice. *Parasite Immunology, 32,* 440–449.

Descoteaux, A., & Turco, S. J. (1999). Glycoconjugates in *Leishmania* infectivity. *Biochimica et Biophysica Acta, 1455,* 341–352.

Desjeux, P. (2004). Leishmaniasis: current situation and new perspectives. Comparative Immunology. *Microbiology and Infectious Diseases, 27,* 305–318.

Diefenbach, A., Schindler, H., Donhauser, N., Lorenz, E., Laskay, T., MacMicking, J., et al. (1998). Type 1 interferon (IFNalpha/beta) and type 2 nitric oxide synthase regulate the innate immune response to a protozoan parasite. *Immunity, 8,* 77–87.

Erb, K., Blank, C., Ritter, U., Bluethmann, H., & Moll, H. (1996). *Leishmania major* infection in major histocompatibility complex class II-deficient mice: CD8+ T cells do not mediate a protective immune response. *Immunobiology, 195,* 243–260.

Faria, M. S., Reis, F. C., Azevedo-Pereira, R. L., Morrison, L. S., Mottram, J. C., & Lima, A. P. (2011). *Leishmania* inhibitor of serine peptidase 2 prevents TLR4 activation by neutrophil elastase promoting parasite survival in murine macrophages. *Journal of Immunology, 186,* 411–422.

Faria, M. S., Reis, F. C., & Lima, A. P. (2012). Toll-like receptors in *Leishmania* infections: guardians or promoters? *Journal of Parasitology Research, 2012,* 930257.

Filardy, A. A., Pires, D. R., Nunes, M. P., Takiya, C. M., Freire-de-Lima, C. G., Ribeiro-Gomes, F. L., et al. (2010). Proinflammatory clearance of apoptotic neutrophils induces an IL-12(low)IL-10(high) regulatory phenotype in macrophages. *Journal of Immunology, 185,* 2044–2050.

Flandin, J. F., Chano, F., & Descoteaux, A. (2006). RNA interference reveals a role for TLR2 and TLR3 in the recognition of *Leishmania donovani* promastigotes by interferon-gamma-primed macrophages. *European Journal of Immunology, 36,* 411–420.

Forget, G., Gregory, D. J., & Olivier, M. (2005). Proteasome-mediated degradation of STAT-1alpha following infection of macrophages with *Leishmania donovani*. *Journal of Biological Chemistry, 280*, 30542–30549.

Forget, G., Gregory, D. J., Whitcombe, L. A., & Olivier, M. (2006). Role of host protein tyrosine phosphatase SHP-1 in *Leishmania donovani*-induced inhibition of nitric oxide production. *Infection and Immunity, 74*, 6272–6279.

de Freitas Balanco, J. M., Moreira, M. E., Bonomo, A., Bozza, P. T., Amarante-Mendes, G., Pirmez, C., et al. (2001). Apoptotic mimicry by an obligate intracellular parasite downregulates macrophage microbicidal activity. *Current Biology, 11*, 1870–1873.

Fruth, U., Solioz, N., & Louis, J. A. (1993). *Leishmania major* interferes with antigen presentation by infected macrophages. *Journal of Immunology, 150*, 1857–1864.

Ganguly, S., Mukhopadhyay, D., Das, N. K., Chaduvula, M., Sadhu, S., Chatterjee, U., et al. (2010). Enhanced lesional Foxp3 expression and peripheral anergic lymphocytes indicate a role for regulatory T cells in Indian post-kala-azar dermal leishmaniasis. *Journal of Investigative Dermatology, 130*, 1013–1022.

Gaur, U., Roberts, S. C., Dalvi, R. P., Corraliza, I., Ullman, B., & Wilson, M. E. (2007). An effect of parasite-encoded arginase on the outcome of murine cutaneous leishmaniasis. *Journal of Immunology, 179*, 8446–8453.

Ghosh, S., Bhattacharyya, S., Das, S., Raha, S., Maulik, N., Das, D. K., et al. (2001). Generation of ceramide in murine macrophages infected with *Leishmania donovani* alters macrophage signaling events and aids intracellular parasitic survival. *Molecular and Cellular Biochemistry, 223*, 47–60.

Ghosh, S., Bhattacharyya, S., Sirkar, M., Sa, G. S., Das, T., Majumdar, D., et al. (2002). *Leishmania donovani* suppresses activated protein 1 and NF-kappaB activation in host macrophages via ceramide generation: involvement of extracellular signal-regulated kinase. *Infection and Immunity, 70*, 6828–6838.

Gomes-Pereira, S., Rodrigues, O. R., Rolão, N., Almeida, P. D., & Santos-Gomes, G. M. (2004). Hepatic cellular immune responses in mice with "cure" and "non-cure" phenotype to *Leishmania infantum* infection: importance of CD8+ T cells and TGF-beta production. *FEMS Immunology and Medical Microbiology, 41*, 59–68.

Gregory, D. J., Godbout, M., Contreras, I., Forget, G., & Olivier, M. (2008). A novel form of NF-kappaB is induced by *Leishmania* infection: involvement in macrophage gene expression. *European Journal of Immunology, 38*, 1071–1081.

Guimarães-Costa, A. B., Nascimento, M. T., Froment, G. S., Soares, R. P., Morgado, F. N., Conceição-Silva, F., et al. (2009). *Leishmania amazonensis* promastigotes induce and are killed by neutrophil extracellular traps. *Proceedings of the National Academy of Sciences of the United States of America, 106*, 6748–6753.

Guizani-Tabbane, L., Ben-Aissa, K., Belghith, M., Sassi, A., & Dellagi, K. (2004). *Leishmania major* amastigotes induce p50/c-Rel NF-kappa B transcription factor in human macrophages: involvement in cytokine synthesis. *Infection and Immunity, 72*, 2582–2589.

Gupta, G., Majumdar, S., Adhikari, A., Bhattacharya, P., Mukherjee, A. K., & Majumdar, S. B. (2011). Treatment with IP-10 induces host-protective immune response by regulating the T regulatory cell functioning in *Leishmania donovani*-infected mice. *Medical Microbiology and Immunology, 200*, 241–253.

Gupta, S., Sane, S. A., Shakya, N., Vishwakarma, P., & Haq, W. (2011). CpG oligodeoxynucleotide 2006 and miltefosine, a potential combination for treatment of experimental visceral leishmaniasis. *Antimicrobial Agents and Chemotherapy, 55*, 3461–3464.

Guy, R. A., & Belosevic, M. (1993). Comparison of receptors required for entry of *Leishmania major* amastigotes into macrophages. *Infection and Immunity, 61*, 1553–1558.

Hartley, M. A., Ronet, C., Zangger, H., Beverley, S. M., & Fasel, N. (2012). *Leishmania* RNA virus: when the host pays the toll. *Frontiers in Cellular and Infection Microbiology, 2*, 99.

Hermoso, T., Fishelson, Z., Becker, S. I., Hirschberg, K., & Jaffe, C. L. (1991). Leishmanial protein kinases phosphorylate components of the complement system. *EMBO Journal, 10*, 4061–4067.

Holm, A., Tejle, K., Magnusson, K. E., Descoteaux, A., & Rasmusson, B. (2001). *Leishmania donovani* lipophosphoglycan causes periphagosomal actin accumulation: correlation with impaired translocation of PKCalpha and defective phagosome maturation. *Cellular Microbiology, 3*, 439–447.

Hoover, D. L., Berger, M., Nacy, C. A., Hockmeyer, W. T., & Meltzer, M. S. (1984). Killing of *Leishmania tropica* amastigotes by factors in normal human serum. *Journal of Immunology, 132*, 893–897.

Huber, M., Timms, E., Mak, T. W., Röllinghoff, M., & Lohoff, M. (1998). Effective and long-lasting immunity against the parasite *Leishmania major* in CD8-deficient mice. *Infection and Immunity, 66*, 3968–3970.

Huynh, C., & Andrews, N. W. (2008). Iron acquisition within host cells and the pathogenicity of *Leishmania*. *Cellular Microbiology, 10*, 293–300.

Iniesta, V., Gómez-Nieto, L. C., & Corraliza, I. (2001). The inhibition of arginase by N(omega)-hydroxy-l-arginine controls the growth of *Leishmania* inside macrophages. *Journal of Experimental Medicine, 193*, 777–784.

Ishikawa, H., Hisaeda, H., Taniguchi, M., Nakayama, T., Sakai, T., Maekawa, Y., et al. (2000). CD4(+) v(alpha)14 NKT cells play a crucial role in an early stage of protective immunity against infection with *Leishmania major*. *International Immunology, 12*, 1267–1274.

Ives, A., Ronet, C., Prevel, F., Ruzzante, G., Fuertes-Marraco, S., Schutz, F., et al. (2011). *Leishmania* RNA virus controls the severity of mucocutaneous leishmaniasis. *Science, 331*, 775–778.

Jaramillo, M., Gomez, M. A., Larsson, O., Shio, M. T., Topisirovic, I., Contreras, I., et al. (2011). *Leishmania* repression of host translation through mTOR cleavage is required for parasite survival and infection. *Cell Host and Microbe, 9*, 331–341.

Kamanaka, M., Yu, P., Yasui, T., Yoshida, K., Kawabe, T., Horii, T., et al. (1996). Protective role of CD40 in *Leishmania major* infection at two distinct phases of cell-mediated immunity. *Immunity, 4*, 275–281.

Kamir, D., Zierow, S., Leng, L., Cho, Y., Diaz, Y., Griffith, J., et al. (2008). A *Leishmania* ortholog of macrophage migration inhibitory factor modulates host macrophage responses. *Journal of Immunology, 180*, 8250–8261.

Kane, M. M., & Mosser, D. M. (2000). *Leishmania* parasites and their ploys to disrupt macrophage activation. *Current Opinion in Hematology, 7*, 26–31.

Karmakar, S., Bhaumik, S. K., Paul, J., & De, T. (2012). TLR4 and NKT cell synergy in immunotherapy against visceral leishmaniasis. *PLoS Pathogens, 8*, e1002646.

Karmakar, S., Paul, J., & De, T. (2011). *Leishmania donovani* glycosphingolipid facilitates antigen presentation by inducing relocation of CD1d into lipid rafts in infected macrophages. *European Journal of Immunology, 41*, 1376–1387.

Katara, G. K., Ansari, N. A., Verma, S., Ramesh, V., & Salotra, P. (2011). Foxp3 and IL-10 expression correlates with parasite burden in lesional tissues of post kala azar dermal leishmaniasis (PKDL) patients. *PLoS Neglected Tropical Diseases, 5*, e1171.

Katzman, S. D., & Fowell, D. J. (2008). Pathogen-imposed skewing of mouse chemokine and cytokine expression at the infected tissue site. *Journal of Clinical Investigation, 118*, 801–811.

Kautz-Neu, K., Noordegraaf, M., Dinges, S., Bennett, C. L., John, D., Clausen, B. E., et al. (2011). Langerhans cells are negative regulators of the anti-*Leishmania* response. *Journal of Experimental Medicine, 208*, 885–891.

Kavoosi, G., Ardestani, S. K., & Kariminia, A. (2009). The involvement of TLR2 in cytokine and reactive oxygen species (ROS) production by PBMCs in response to *Leishmania major* phosphoglycans (PGs). *Parasitology, 136*, 1193–1199.

Kavoosi, G., Ardestani, S. K., Kariminia, A., & Alimohammadian, M. H. (2010). *Leishmania major* lipophosphoglycan: discrepancy in toll-like receptor signaling. *Experimental Parasitology, 124,* 214–218.

Kaye, P., & Scott, P. (2011). Leishmaniasis: complexity at the host-pathogen interface. *Nature Reviews Microbiology, 9,* 604–615.

Khouri, R., Bafica, A., Silva, M. a.P., Noronha, A., Kolb, J. P., Wietzerbin, J., et al. (2009). IFN-beta impairs superoxide-dependent parasite killing in human macrophages: evidence for a deleterious role of SOD1 in cutaneous leishmaniasis. *Journal of Immunology, 182,* 2525–2531.

Kima, P. E., Soong, L., Chicharro, C., Ruddle, N. H., & McMahon-Pratt, D. (1996). *Leishmania*-infected macrophages sequester endogenously synthesized parasite antigens from presentation to CD4+ T cells. *European Journal of Immunology, 26,* 3163–3169.

Kirkpatrick, C. E., & Farrell, J. P. (1982). Leishmaniasis in beige mice. *Infection and Immunity, 38,* 1208–1216.

Kopf, M., Brombacher, F., Kohler, G., Kienzle, G., Widmann, K. H., Lefrang, K., et al. (1996). IL-4-deficient Balb/c mice resist infection with *Leishmania major. Journal of Experimental Medicine, 184,* 1127–1136.

Kropf, P., Freudenberg, N., Kalis, C., Modolell, M., Herath, S., Galanos, C., et al. (2004). Infection of C57BL/10ScCr and C57BL/10ScNCr mice with *Leishmania major* reveals a role for toll-like receptor 4 in the control of parasite replication. *Journal of Leukocyte Biology, 76,* 48–57.

Kropf, P., Freudenberg, M. A., Modolell, M., Price, H. P., Herath, S., Antoniazi, S., et al. (2004). Toll-like receptor 4 contributes to efficient control of infection with the protozoan parasite *Leishmania major. Infection and Immunity, 72,* 1920–1928.

Kropf, P., Herath, S., Weber, V., Modolell, M., & Müller, I. (2003). Factors influencing *Leishmania major* infection in IL-4-deficient BALB/c mice. *Parasite Immunology, 25,* 439–447.

Kushawaha, P. K., Gupta, R., Sundar, S., Sahasrabuddhe, A. A., & Dube, A. (2011). Elongation factor-2, a Th1 stimulatory protein of *Leishmania donovani*, generates strong IFN-γ and IL-12 response in cured *Leishmania*-infected patients/hamsters and protects hamsters against *Leishmania* challenge. *Journal of Immunology, 187,* 6417–6427.

Lang, T., de Chastellier, C., Frehel, C., Hellio, R., Metezeau, P., Leao, S. e.S., et al. (1994). Distribution of MHC class I and of MHC class II molecules in macrophages infected with *Leishmania amazonensis. Journal of Cell Science, 107*(Pt 1), 69–82.

Laskay, T., Diefenbach, A., Röllinghoff, M., & Solbach, W. (1995). Early parasite containment is decisive for resistance to *Leishmania major* infection. *European Journal of Immunology, 25,* 2220–2227.

Lieke, T., Nylén, S., Eidsmo, L., Schmetz, C., Berg, L., & Akuffo, H. (2011). The interplay between *Leishmania* promastigotes and human natural killer cells in vitro leads to direct lysis of *Leishmania* by NK cells and modulation of NK cell activity by *Leishmania* promastigotes. *Parasitology, 138,* 1898–1909.

Liu, D., Zhang, T., Marshall, A. J., Okkenhaug, K., Vanhaesebroeck, B., & Uzonna, J. E. (2009). The p110delta isoform of phosphatidylinositol 3-kinase controls susceptibility to *Leishmania major* by regulating expansion and tissue homing of regulatory T cells. *Journal of Immunology, 183,* 1921–1933.

Llanos-Cuentas, A., Calderon, W., Cruz, M., Ashman, J. A., Alves, F. P., Coler, R. N., et al. (2010). A clinical trial to evaluate the safety and immunogenicity of the LEISH-F1+MPL-SE vaccine when used in combination with sodium stibogluconate for the treatment of mucosal leishmaniasis. *Vaccine, 28,* 7427–7435.

Lo, S. K., Bovis, L., Matura, R., Zhu, B., He, S., Lum, H., et al. (1998). *Leishmania* lipophosphoglycan reduces monocyte transendothelial migration: modulation of cell adhesion molecules, intercellular junctional proteins, and chemoattractants. *Journal of Immunology, 160,* 1857–1865.

Locksley, R. M., Reiner, S. L., Hatam, F., Littman, D. R., & Killeen, N. (1993). Helper T cells without CD4: control of leishmaniasis in CD4-deficient mice. *Science, 261*, 1448–1451.

Mannheimer, S. B., Hariprashad, J., Stoeckle, M. Y., & Murray, H. W. (1996). Induction of macrophage antiprotozoal activity by monocyte chemotactic and activating factor. *FEMS Immunology and Medical Microbiology, 14*, 59–61.

Marth, T., & Kelsall, B. L. (1997). Regulation of interleukin-12 by complement receptor 3 signaling. *Journal of Experimental Medicine, 185*, 1987–1995.

Martiny, A., Meyer-Fernandes, J. R., de Souza, W., & Vannier-Santos, M. A. (1999). Altered tyrosine phosphorylation of ERK1 MAP kinase and other macrophage molecules caused by *Leishmania* amastigotes. *Molecular and Biochemical Parasitology, 102*, 1–12.

Mary, C., Auriault, V., Faugère, B., & Dessein, A. J. (1999). Control of *Leishmania infantum* infection is associated with CD8(+) and gamma interferon- and interleukin-5-producing CD4(+) antigen-specific T cells. *Infection and Immunity, 67*, 5559–5566.

Mathur, R. K., Awasthi, A., Wadhone, P., Ramanamurthy, B., & Saha, B. (2004). Reciprocal CD40 signals through p38MAPK and ERK-1/2 induce counteracting immune responses. *Nature Medicine, 10*, 540–544.

Mattner, J., Donhauser, N., Werner-Felmayer, G., & Bogdan, C. (2006). NKT cells mediate organ-specific resistance against *Leishmania major* infection. *Microbes and Infection, 8*, 354–362.

Mattner, F., Magram, J., Ferrante, J., Launois, P., Di Padova, K., Behin, R., et al. (1996). Genetically resistant mice lacking interleukin-12 are susceptible to infection with *Leishmania major* and mount a polarized Th2 cell response. *European Journal of Immunology, 26*, 1553–1559.

Mbow, M. L., DeKrey, G. K., & Titus, R. G. (2001). *Leishmania major* induces differential expression of costimulatory molecules on mouse epidermal cells. *European Journal of Immunology, 31*, 1400–1409.

McDowell, M. A., & Sacks, D. L. (1999). Inhibition of host cell signal transduction by *Leishmania*: observations relevant to the selective impairment of IL-12 responses. *Current Opinion in Microbiology, 2*, 438–443.

Meddeb-Garnaoui, A., Zrelli, H., & Dellagi, K. (2009). Effects of tropism and virulence of *Leishmania* parasites on cytokine production by infected human monocytes. *Clinical and Experimental Immunology, 155*, 199–206.

Mendez, S., Reckling, S. K., Piccirillo, C. A., Sacks, D., & Belkaid, Y. (2004). Role for CD4(+) CD25(+) regulatory T cells in reactivation of persistent leishmaniasis and control of concomitant immunity. *Journal of Experimental Medicine, 200*, 201–210.

Mise-Omata, S., Kuroda, E., Sugiura, T., Yamashita, U., Obata, Y., & Doi, T. S. (2009). The NF-kappaB RelA subunit confers resistance to *Leishmania major* by inducing nitric oxide synthase 2 and Fas expression but not Th1 differentiation. *Journal of Immunology, 182*, 4910–4916.

Mollinedo, F., Janssen, H., de la Iglesia-Vicente, J., Villa-Pulgarin, J. A., & Calafat, J. (2010). Selective fusion of azurophilic granules with *Leishmania*-containing phagosomes in human neutrophils. *Journal of Biological Chemistry, 285*, 34528–34536.

Moreno, I., Molina, R., Toraño, A., Laurin, E., García, E., & Domínguez, M. (2007). Comparative real-time kinetic analysis of human complement killing of *Leishmania infantum* promastigotes derived from axenic culture or from Phlebotomus perniciosus. *Microbes and Infection, 9*, 1574–1580.

Mosser, D. M., & Edelson, P. J. (1985). The mouse macrophage receptor for C3bi (CR3) is a major mechanism in the phagocytosis of *Leishmania* promastigotes. *Journal of Immunology, 135*, 2785–2789.

Müller, K., van Zandbergen, G., Hansen, B., Laufs, H., Jahnke, N., Solbach, W., et al. (2001). Chemokines, natural killer cells and granulocytes in the early course of *Leishmania major* infection in mice. *Medical Microbiology and Immunology, 190*, 73–76.

Murphy, M. L., Cotterell, S. E., Gorak, P. M., Engwerda, C. R., & Kaye, P. M. (1998). Blockade of CTLA-4 enhances host resistance to the intracellular pathogen, *Leishmania donovani*. *Journal of Immunology, 161,* 4153–4160.

Murphy, M. L., Engwerda, C. R., Gorak, P. M., & Kaye, P. M. (1997). B7-2 blockade enhances T cell responses to *Leishmania donovani*. *Journal of Immunology, 159,* 4460–4466.

Murray, H. W., & Nathan, C. F. (1999). Macrophage microbicidal mechanisms in vivo: reactive nitrogen versus oxygen intermediates in the killing of intracellular visceral *Leishmania donovani*. *Journal of Experimental Medicine, 189,* 741–746.

Nandan, D., Lo, R., & Reiner, N. E. (1999). Activation of phosphotyrosine phosphatase activity attenuates mitogen-activated protein kinase signaling and inhibits c-FOS and nitric oxide synthase expression in macrophages infected with *Leishmania donovani*. *Infection and Immunity, 67,* 4055–4063.

Nandan, D., & Reiner, N. E. (1995). Attenuation of gamma interferon-induced tyrosine phosphorylation in mononuclear phagocytes infected with *Leishmania donovani*: selective inhibition of signaling through Janus kinases and Stat1. *Infection and Immunity, 63,* 4495–4500.

Nateghi Rostami, M., Keshavarz, H., Edalat, R., Sarrafnejad, A., Shahrestani, T., Mahboudi, F., et al. (2010). CD8+ T cells as a source of IFN-γ production in human cutaneous leishmaniasis. *PLoS Neglected Tropical Diseases, 4,* e845.

Novais, F. O., Santiago, R. C., Báfica, A., Khouri, R., Afonso, L., Borges, V. M., et al. (2009). Neutrophils and macrophages cooperate in host resistance against *Leishmania braziliensis* infection. *Journal of Immunology, 183,* 8088–8098.

Oghumu, S., Lezama-Dávila, C. M., Isaac-Márquez, A. P., & Satoskar, A. R. (2010). Role of chemokines in regulation of immunity against leishmaniasis. *Experimental Parasitology, 126,* 389–396.

Olivier, M., Atayde, V. D., Isnard, A., Hassani, K., & Shio, M. T. (2012). *Leishmania* virulence factors: focus on the metalloprotease GP63. *Microbes and Infection, 14,* 1377–1389.

Olivier, M., Brownsey, R. W., & Reiner, N. E. (1992). Defective stimulus-response coupling in human monocytes infected with *Leishmania donovani* is associated with altered activation and translocation of protein kinase C. *Proceedings of the National Academy of Sciences of the United States of America, 89,* 7481–7485.

Olivier, M., Gregory, D. J., & Forget, G. (2005). Subversion mechanisms by which *Leishmania* parasites can escape the host immune response: a signaling point of view. *Clinical Microbiology Reviews, 18,* 293–305.

Pearson, R. D., & Steigbigel, R. T. (1981). Phagocytosis and killing of the protozoan *Leishmania donovani* by human polymorphonuclear leukocytes. *Journal of Immunology, 127,* 1438–1443.

Pereira, L. I., Dorta, M. L., Pereira, A. J., Bastos, R. P., Oliveira, M. A., Pinto, S. A., et al. (2009). Increase of NK cells and proinflammatory monocytes are associated with the clinical improvement of diffuse cutaneous leishmaniasis after immunochemotherapy with BCG/*Leishmania* antigens. *American Journal of Tropical Medicine and Hygiene, 81,* 378–383.

Pereira, W. F., Ribeiro-Gomes, F. L., Guillermo, L. V., Vellozo, N. S., Montalvão, F., Dosreis, G. A., et al. (2011). Myeloid-derived suppressor cells help protective immunity to *Leishmania major* infection despite suppressed T cell responses. *Journal of Leukocyte Biology, 90,* 1191–1197.

Peruhype-Magalhães, V., Martins-Filho, O. A., Prata, A., Silva, L. e.A., Rabello, A., Teixeira-Carvalho, A., et al. (2005). Immune response in human visceral leishmaniasis: analysis of the correlation between innate immunity cytokine profile and disease outcome. *Scandinavian Journal of Immunology, 62,* 487–495.

Peters, N., & Sacks, D. (2006). Immune privilege in sites of chronic infection: leishmania and regulatory T cells. *Immunological Reviews, 213,* 159–179.

Pinheiro, N. F., Hermida, M. D., Macedo, M. P., Mengel, J., Bafica, A., & dos-Santos, W. L. (2006). *Leishmania* infection impairs beta 1-integrin function and chemokine receptor expression in mononuclear phagocytes. *Infection and Immunity, 74,* 3912–3921.

Pompeu, M. L., Freitas, L. A., Santos, M. L., Khouri, M., & Barral-Netto, M. (1991). Granulocytes in the inflammatory process of BALB/c mice infected by *Leishmania amazonensis*. A quantitative approach. *Acta Tropica, 48*, 185–193.

Prajeeth, C. K., Haeberlein, S., Sebald, H., Schleicher, U., & Bogdan, C. (2011). *Leishmania*-infected macrophages are targets of NK cell-derived cytokines but not of NK cell cytotoxicity. *Infection and Immunity, 79*, 2699–2708.

Prina, E., Lang, T., Glaichenhaus, N., & Antoine, J. C. (1996). Presentation of the protective parasite antigen LACK by *Leishmania*-infected macrophages. *Journal of Immunology, 156*, 4318–4327.

Privé, C., & Descoteaux, A. (2000). *Leishmania donovani* promastigotes evade the activation of mitogen-activated protein kinases p38, c-Jun N-terminal kinase, and extracellular signal-regulated kinase-1/2 during infection of naive macrophages. *European Journal of Immunology, 30*, 2235–2244.

Puentes, S. M., Da Silva, R. P., Sacks, D. L., Hammer, C. H., & Joiner, K. A. (1990). Serum resistance of metacyclic stage *Leishmania major* promastigotes is due to release of C5b-9. *Journal of Immunology, 145*, 4311–4316.

Puentes, S. M., Sacks, D. L., da Silva, R. P., & Joiner, K. A. (1988). Complement binding by two developmental stages of *Leishmania major* promastigotes varying in expression of a surface lipophosphoglycan. *Journal of Experimental Medicine, 167*, 887–902.

Rai, A. K., Thakur, C. P., Seth, T., & Mitra, D. K. (2011). Enrichment of invariant natural killer T cells in the bone marrow of visceral leishmaniasis patients. *Parasite Immunology, 33*, 688–691.

Rai, A. K., Thakur, C. P., Singh, A., Seth, T., Srivastava, S. K., Singh, P., et al. (2012). Regulatory T cells suppress T cell activation at the pathologic site of human visceral leishmaniasis. *PLoS One, 7*, e31551.

Raman, V. S., Bhatia, A., Picone, A., Whittle, J., Bailor, H. R., O'Donnell, J., et al. (2010). Applying TLR synergy in immunotherapy: implications in cutaneous leishmaniasis. *Journal of Immunology, 185*, 1701–1710.

Reguera, R. M., Balaña-Fouce, R., Showalter, M., Hickerson, S., & Beverley, S. M. (2009). *Leishmania major* lacking arginase (ARG) are auxotrophic for polyamines but retain infectivity to susceptible BALB/c mice. *Molecular and Biochemical Parasitology, 165*, 48–56.

Reiner, S. L., Zheng, S., Wang, Z. E., Stowring, L., & Locksley, R. M. (1994). *Leishmania* promastigotes evade interleukin 12 (IL-12) induction by macrophages and stimulate a broad range of cytokines from CD4+ T cells during initiation of infection. *Journal of Experimental Medicine, 179*, 447–456.

Rezai, H. R., Sher, S., & Gettner, S. (1969). *Leishmania tropica, L. donovani*, and *L. enriettii*: immune rabbit serum inhibitory in vitro. *Experimental Parasitology, 26*, 257–263.

Ribeiro-Gomes, F. L., Moniz-de-Souza, M. C., Alexandre-Moreira, M. S., Dias, W. B., Lopes, M. F., Nunes, M. P., et al. (2007). Neutrophils activate macrophages for intracellular killing of *Leishmania major* through recruitment of TLR4 by neutrophil elastase. *Journal of Immunology, 179*, 3988–3994.

Ribeiro-Gomes, F. L., Otero, A. C., Gomes, N. A., Moniz-De-Souza, M. C., Cysne-Finkelstein, L., Arnholdt, A. C., et al. (2004). Macrophage interactions with neutrophils regulate *Leishmania major* infection. *Journal of Immunology, 172*, 4454–4462.

Ritter, U., Moll, H., Laskay, T., Bröcker, E., Velazco, O., Becker, I., et al. (1996). Differential expression of chemokines in patients with localized and diffuse cutaneous American leishmaniasis. *Journal of Infectious Diseases, 173*, 699–709.

Sacks, D. L., Pimenta, P. F., McConville, M. J., Schneider, P., & Turco, S. J. (1995). Stage-specific binding of *Leishmania donovani* to the sand fly vector midgut is regulated by conformational changes in the abundant surface lipophosphoglycan. *Journal of Experimental Medicine, 181*, 685–697.

Saha, B., Das, G., Vohra, H., Ganguly, N. K., & Mishra, G. C. (1995). Macrophage-T cell interaction in experimental visceral leishmaniasis: failure to express costimulatory molecules on *Leishmania*-infected macrophages and its implication in the suppression of cell-mediated immunity. *European Journal of Immunology, 25*, 2492–2498.

Satoskar, A. R., Stamm, L. M., Zhang, X., Okano, M., David, J. R., Terhorst, C., et al. (1999). NK cell-deficient mice develop a Th1-like response but fail to mount an efficient antigen-specific IgG2a antibody response. *Journal of Immunology, 163*, 5298–5302.

Scharton, T. M., & Scott, P. (1993). Natural killer cells are a source of interferon gamma that drives differentiation of CD4+ T cell subsets and induces early resistance to *Leishmania major* in mice. *Journal of Experimental Medicine, 178*, 567–577.

Scianimanico, S., Desrosiers, M., Dermine, J. F., Méresse, S., Descoteaux, A., & Desjardins, M. (1999). Impaired recruitment of the small GTPase rab7 correlates with the inhibition of phagosome maturation by *Leishmania donovani* promastigotes. *Cellular Microbiology, 1*, 19–32.

Sen, S., Roy, K., Mukherjee, S., Mukhopadhyay, R., & Roy, S. (2011). Restoration of IFNγR subunit assembly, IFNγ signaling and parasite clearance in *Leishmania donovani* infected macrophages: role of membrane cholesterol. *PLoS Pathogens, 7*, e1002229.

Shakya, N., Sane, S. A., Shankar, S., & Gupta, S. (2011). Effect of Pam3Cys induced protection on the therapeutic efficacy of miltefosine against experimental visceral leishmaniasis. *Peptides, 32*, 2131–2133.

Shweash, M., Adrienne McGachy, H., Schroeder, J., Neamatallah, T., Bryant, C. E., Millington, O., et al. (2011). *Leishmania mexicana* promastigotes inhibit macrophage IL-12 production via TLR-4 dependent COX-2, iNOS and arginase-1 expression. *Molecular Immunology, 48*, 1800–1808.

Silva, V. M., Larangeira, D. F., Oliveira, P. R., Sampaio, R. B., Suzart, P., Nihei, J. S., et al. (2011). Enhancement of experimental cutaneous leishmaniasis by *Leishmania* molecules is dependent on interleukin-4, serine protease/esterase activity, and parasite and host genetic backgrounds. *Infection and Immunity, 79*, 1236–1243.

Smelt, S. C., Cotterell, S. E., Engwerda, C. R., & Kaye, P. M. (2000). B cell-deficient mice are highly resistant to *Leishmania donovani* infection, but develop neutrophil-mediated tissue pathology. *Journal of Immunology, 164*, 3681–3688.

de Souza Carmo, E. V., Katz, S., & Barbiéri, C. L. (2010). Neutrophils reduce the parasite burden in *Leishmania* (*Leishmania*) *amazonensis*-infected macrophages. *PLoS One, 5*, e13815.

Srivastava, A., Singh, N., Mishra, M., Kumar, V., Gour, J. K., Bajpai, S., et al. (2012). Identification of TLR inducing Th1-responsive *Leishmania donovani* amastigote-specific antigens. *Molecular and Cellular Biochemistry, 359*, 359–368.

Srivastava, N., Sudan, R., & Saha, B. (2011). CD40-modulated dual-specificity phosphatases MAPK phosphatase (MKP)-1 and MKP-3 reciprocally regulate *Leishmania major* infection. *Journal of Immunology, 186*, 5863–5872.

Srivastav, S., Kar, S., Chande, A. G., Mukhopadhyaya, R., & Das, P. K. (2012). *Leishmania donovani* exploits host deubiquitinating enzyme A20, a negative regulator of TLR signaling, to subvert host immune response. *Journal of Immunology, 189*, 924–934.

Stanley, A. C., Zhou, Y., Amante, F. H., Randall, L. M., Haque, A., Pellicci, D. G., et al. (2008). Activation of invariant NKT cells exacerbates experimental visceral leishmaniasis. *PLoS Pathogens, 4*, e1000028.

Suffia, I., Reckling, S. K., Salay, G., & Belkaid, Y. (2005). A role for CD103 in the retention of CD4+CD25+ Treg and control of *Leishmania major* infection. *Journal of Immunology, 174*, 5444–5455.

Suzuki, E., Tanaka, A. K., Toledo, M. S., Takahashi, H. K., & Straus, A. H. (2002). Role of beta-D-galactofuranose in *Leishmania major* macrophage invasion. *Infection and Immunity, 70*, 6592–6596.

Swihart, K., Fruth, U., Messmer, N., Hug, K., Behin, R., Huang, S., et al. (1995). Mice from a genetically resistant background lacking the interferon gamma receptor are susceptible to infection with *Leishmania major* but mount a polarized T helper cell 1-type CD4+ T cell response. *Journal of Experimental Medicine, 181*, 961–971.

Tabatabaee, P. A., Abolhassani, M., Mahdavi, M., Nahrevanian, H., & Azadmanesh, K. (2011). *Leishmania major*: secreted antigens of *Leishmania major* promastigotes shift the immune response of the C57BL/6 mice toward Th2 in vitro. *Experimental Parasitology, 127*, 46–51.

Tacchini-Cottier, F., Zweifel, C., Belkaid, Y., Mukankundiye, C., Vasei, M., Launois, P., et al. (2000). An immunomodulatory function for neutrophils during the induction of a CD4+ Th2 response in BALB/c mice infected with *Leishmania major*. *Journal of Immunology, 165*, 2628–2636.

Talamás-Rohana, P., Wright, S. D., Lennartz, M. R., & Russell, D. G. (1990). Lipophosphoglycan from *Leishmania mexicana* promastigotes binds to members of the CR3, p150,95 and LFA-1 family of leukocyte integrins. *Journal of Immunology, 144*, 4817–4824.

Thalhofer, C. J., Chen, Y., Sudan, B., Love-Homan, L., & Wilson, M. E. (2011). Leukocytes infiltrate the skin and draining lymph nodes in response to the protozoan *Leishmania infantum chagasi*. *Infection and Immunity, 79*, 108–117.

Tsagozis, P., Karagouni, E., & Dotsika, E. (2005). Function of CD8+ T lymphocytes in a self-curing mouse model of visceral leishmaniasis. *Parasitology International, 54*, 139–146.

Turco, S. J., Hull, S. R., Orlandi, P. A., Shepherd, S. D., Homans, S. W., Dwek, R. A., et al. (1987). Structure of the major carbohydrate fragment of the *Leishmania donovani* lipophosphoglycan. *Biochemistry, 26*, 6233–6238.

Vargas-Inchaustegui, D. A., Tai, W., Xin, L., Hogg, A. E., Corry, D. B., & Soong, L. (2009). Distinct roles for MyD88 and toll-like receptor 2 during *Leishmania braziliensis* infection in mice. *Infection and Immunity, 77*, 2948–2956.

de Veer, M. J., Curtis, J. M., Baldwin, T. M., DiDonato, J. A., Sexton, A., McConville, M. J., et al. (2003). MyD88 is essential for clearance of *Leishmania major*: possible role for lipophosphoglycan and toll-like receptor 2 signaling. *European Journal of Immunology, 33*, 2822–2831.

Vinet, A. F., Fukuda, M., Turco, S. J., & Descoteaux, A. (2009). The *Leishmania donovani* lipophosphoglycan excludes the vesicular proton-ATPase from phagosomes by impairing the recruitment of synaptotagmin V. *PLoS Pathogens, 5*, e1000628.

Vivarini, A. e. C., Pereira, R. e. M., Teixeira, K. L., Calegari-Silva, T. C., Bellio, M., Laurenti, M. D., et al. (2011). Human cutaneous leishmaniasis: interferon-dependent expression of double-stranded RNA-dependent protein kinase (PKR) via TLR2. *FASEB Journal, 25*, 4162–4173.

Weingartner, A., Kemmer, G., Muller, F. D., Zampieri, R. A., Gonzaga Dos Santos, M., Schiller, J., et al. (2012). *Leishmania* promastigotes lack phosphatidylserine but bind annexin V upon permeabilization or miltefosine treatment. *PLoS One, 7*, e42070.

Weinheber, N., Wolfram, M., Harbecke, D., & Aebischer, T. (1998). Phagocytosis of *Leishmania mexicana* amastigotes by macrophages leads to a sustained suppression of IL-12 production. *European Journal of Immunology, 28*, 2467–2477.

Whitaker, S. M., Colmenares, M., Pestana, K. G., & McMahon-Pratt, D. (2008). *Leishmania pifanoi* proteoglycolipid complex P8 induces macrophage cytokine production through toll-like receptor 4. *Infection and Immunity, 76*, 2149–2156.

Wiethe, C., Debus, A., Mohrs, M., Steinkasserer, A., Lutz, M., & Gessner, A. (2008). Dendritic cell differentiation state and their interaction with NKT cells determine Th1/Th2 differentiation in the murine model of *Leishmania major* infection. *Journal of Immunology, 180*, 4371–4381.

Wilson, J., Huynh, C., Kennedy, K. A., Ward, D. M., Kaplan, J., Aderem, A., et al. (2008). Control of parasitophorous vacuole expansion by LYST/Beige restricts the intracellular growth of *Leishmania amazonensis*. *PLoS Pathogens, 4,* e1000179.

Wright, S. D., & Silverstein, S. C. (1983). Receptors for C3b and C3bi promote phagocytosis but not the release of toxic oxygen from human phagocytes. *Journal of Experimental Medicine, 158,* 2016–2023.

Yang, Z., Mosser, D. M., & Zhang, X. (2007). Activation of the MAPK, ERK, following *Leishmania amazonensis* infection of macrophages. *Journal of Immunology, 178,* 1077–1085.

Yurchenko, E., Tritt, M., Hay, V., Shevach, E. M., Belkaid, Y., & Piccirillo, C. A. (2006). CCR5-dependent homing of naturally occurring CD4+ regulatory T cells to sites of *Leishmania major* infection favors pathogen persistence. *Journal of Experimental Medicine, 203,* 2451–2460.

van Zandbergen, G., Hermann, N., Laufs, H., Solbach, W., & Laskay, T. (2002). *Leishmania* promastigotes release a granulocyte chemotactic factor and induce interleukin-8 release but inhibit gamma interferon-inducible protein 10 production by neutrophil granulocytes. *Infection and Immunity, 70,* 4177–4184.

INDEX

Note: Page numbers followed by "f" and "t" indicate figures and tables respectively.

CONTENTS OF PREVIOUS VOLUMES

Printed and bound by CPI Group (UK) Ltd, Croydon, CR0 4YY

08/05/2025

01864958-0001